China, the USA and Technological Supremacy in Europe

W0113814

This book explores how technological competition is linked to the geopolitical contest between the United States and China, and why Europe and the European Union (EU) have become involved in this competition for technological supremacy.

China's political and economic rise, the concurrent US withdrawal from the region, and the rise of new technologies such as 5G and AI create a new and more unstable geopolitical environment in the region. In addition, the EU, far from being a global player, finds it increasingly difficult to play a leading role. This book analyzes the nature of the ultimate goal of technological competition between the United States and China and shows how and why did the EU become the center of this struggle. The author argues that the EU has become the new battlefield of the technological struggle since wealthy societies in the EU make this competition attractive and profitable to both the US and China.

By shedding light on the geopolitical motivations of China and the question of whether the US can contain China's advance in this domain, the book will be of interest to practitioners in the fields of international relations and political science as well as policymakers and analysts employed by diplomatic services, multilateral organizations, and nongovernmental organizations.

Csaba Moldicz, PhD, is an associate professor at the Department of International Relations, Faculty of International Business and Management, Budapest Business School, Hungary.

Rethinking Asia and International Relations

Series Editor – Emilian Kavalski, Li Dak Sum Chair Professor in China-Eurasia Relations and International Studies, University of Nottingham, Ningbo, China

This series seeks to provide thoughtful consideration both of the growing prominence of Asian actors on the global stage and the changes in the study and practice of world affairs that they provoke. It intends to offer a comprehensive parallel assessment of the full spectrum of Asian states, organisations, and regions and their impact on the dynamics of global politics.

The series seeks to encourage conversation on:

- what rules, norms, and strategic cultures are likely to dominate international life in the 'Asian Century';
- how will global problems be reframed and addressed by a 'rising Asia';
- which institutions, actors, and states are likely to provide leadership during such 'shifts to the East';
- whether there is something distinctly 'Asian' about the emerging patterns of global politics.

Such comprehensive engagement not only aims to offer a critical assessment of the actual and prospective roles of Asian actors, but also seeks to rethink the concepts, practices, and frameworks of analysis of world politics.

This series invites proposals for interdisciplinary research monographs undertaking comparative studies of Asian actors and their impact on the current patterns and likely future trajectories of international relations. Furthermore, it offers a platform for pioneering explorations of the ongoing transformations in global politics as a result of Asia's increasing centrality to the patterns and practices of world affairs.

China, the USA and Technological Supremacy in Europe
Csaba Moldicz

China and Eurasia
Rethinking Cooperation and Contradictions
in the Era of Changing World Order
Mher D. Sahakyan and Heinz Gärtner

For more information about this series, please visit: https://www.routledge.com/Rethinking-Asia-and-International-Relations/book-series/ASHSER1384

China, the USA and Technological Supremacy in Europe

Csaba Moldicz

Routledge
Taylor & Francis Group

LONDON AND NEW YORK

First published 2022
by Routledge
2 Park Square, Milton Park, Abingdon, Oxon OX14 4RN

and by Routledge
52 Vanderbilt Avenue, New York, NY 10017

Routledge is an imprint of the Taylor & Francis Group, an informa business

© 2022 Csaba Moldicz

The right of Csaba Moldicz to be identified as author of this work
has been asserted by her in accordance with sections 77 and 78 of
the Copyright, Designs and Patents Act 1988.

All rights reserved. No part of this book may be reprinted or
reproduced or utilised in any form or by any electronic, mechanical,
or other means, now known or hereafter invented, including
photocopying and recording, or in any information storage or
retrieval system, without permission in writing from the publishers.

Trademark notice: Product or corporate names may be trademarks
or registered trademarks, and are used only for identification and
explanation without intent to infringe.

British Library Cataloguing-in-Publication Data
A catalogue record for this book is available from the British Library

Library of Congress Cataloging-in-Publication Data
Names: Moldicz, Csaba, author.
Title: China, the United States and the technological supremacy in
Europe / Csaba Moldicz.
Description: Abingdon, Oxon ; New York, NY : Routledge, [2022] |
Series: Rethinking Asia and international relations | Includes
bibliographical references and index.
Identifiers: LCCN 2021008558 (print) | LCCN 2021008559 (ebook) |
ISBN 9780367652500 (hardback) | ISBN 9780367652548 (paperback) |
ISBN 9781003128625 (ebook)
Subjects: LCSH: Technology and state. | Competition, International. |
Geopolitics. | United States—Foreign relations—China. | China—
Foreign relations—United States. | China—Economic conditions—
21st century. | United States—Economic conditions—21st century. |
European Union—Economic conditions—21st century.
Classification: LCC T20 .M55 2022 (print) | LCC T20 (ebook) |
DDC 338.9/26—dc23
LC record available at https://lccn.loc.gov/2021008558
LC ebook record available at https://lccn.loc.gov/2021008559

ISBN: 978-0-367-65250-0 (hbk)
ISBN: 978-0-367-65254-8 (pbk)
ISBN: 978-1-003-12862-5 (ebk)

DOI: 10.4324/9781003128625

Typeset in Times New Roman
by codeMantra

Contents

Tables

Author biography

Csaba Moldicz, PhD, is an associate professor at the Department of International Relations, Faculty of International Business and Management, Budapest Business School, Hungary. His main research area is the economic integration process of the EU and China, with a special focus on the Central and Eastern European region. Currently, he is a research director at Oriental Business and Innovation Centre, which was established in 2016 by the Budapest Business School and the Central Bank of Hungary. He is an associate research fellow of the Institute for Foreign Affairs and Trade (Hungary) and the China-CEE Institute.

Preface

Authors often begin their work by asking about the relevance of the book, its timing, and where this research is taking us. In this case, it is hardly necessary to convince the reader why the topic of *China, the USA and the Technological Supremacy in Europe* is important in our world, and our time; however, disentangling these elements might help the reader understand how and why technological competition is linked to the geopolitical contest between the United States and China, why the European aspect is of greater importance to the outcome than any other region, and how the EU became a battleground in this competition for technological supremacy.

It must be stressed this book was finalized in early February of 2021; thus, it could not include major policy changes in the United States, which seems to be very likely, as the new President—unlike Donald Trump—emphasized the willingness to cooperate with China in its first speech on America's place in the world (Biden, February 4, 2021). The book does not focus on the technical details of this competition either. The main contribution of this work is to contextualize this competition in terms of geopolitics or geoeconomics. The main reason this work centers on technology development, particularly the fifth generation (5G) mobile networks, is that this technology standard for mobile phones is considered a keystone for future economic, political, and security frameworks. The 5G provides the fundamental infrastructure that will enable faster, simpler, and more efficient networks. It is no wonder that political, military, and business decision-makers are desperate to win this race. It may be an exaggeration to say that the country that wins this competition can rule the world, but the superpowers of this new world must also lead in 5G.

There is another aspect of this competition that analysts tend to overlook. The 5G mobile networks along with other new technologies (artificial intelligence, Internet of Things, etc.) are changing the traditional boundaries between the state and its citizens. As technology changes, so does the technology of surveillance and the resulting capabilities that the state apparatus has access to. On the one hand, we tend to argue that civil society can and should enforce respect for privacy in democratic societies, and this is true, but on the other hand, we need to understand that political regimes are also in competition, so these boundaries will shift in democratic societies, too, in favor of less privacy. Thus, the outcome of the struggle between the United States and China is not only the battle of two superpowers, but also it will be a defining element determining our daily lives in the near future.

The next question we can raise about the book is why we have this strong focus on Europe. In our view, the European continent is important in this competition for two reasons. Europe, particularly the EU, is wealthy; it is a major market for new technology. At the same time, the EU is militarily weak, where the formulation of a common foreign policy was a huge problem from the beginning. The first big mistakes made in the framework of Common Foreign and Security Policy (CFSP) go back to the break-up of Yugoslavia in the 1990s, and since then foreign policy cooperation has been a nightmare for European politicians. We do not exaggerate when we say that drafting, adopting, and implementing a coherent China strategy in the EU have run into significant obstacles in recent years. The main reason in this field for failing cooperation among EU member states is simply their different economic and political interests in the cooperation with China. Two chapters systematically analyze these interests and the legal and business frameworks of two groups of countries within the EU: the three largest economies (Germany, France, and Italy) and the Visegrád countries.

The first chapter, "The battle for technology: the global arena," using selected indicators with a particular focus on 5G and artificial intelligence, provides an overview of the struggle for technological supremacy between China and the United States. This chapter endeavors to answer the question of whether or not the accusations that China is secretly seeking technological supremacy can be proven and, more importantly, whether or not this goal can be achieved. In addition to analyzing some of the main indicators of this struggle between the two countries, special attention will be paid to the development of ideas that attempt to frame the rapid growth of the Chinese economy—the "China's rise debate," "the decoupling vs. engagement dilemma" of US foreign policy, and the paradigm of the "Asian developmental state"—and place them in a geopolitical and geoeconomic context. This overview of ideas and indicators will lead us to the second chapter of this book and prepare us to understand what is happening on the European continent.

In the chapter titled "The war of arguments: the European battlefield," we focus on the EU. As we said above, the EU's political leverage in the world is limited, and it is strangled by its own institutional problems which take form in the absence of an effective industrial policy at the EU level. This problem is further complicated by the fragmented market, as 5G and technology- and investment-related areas are in the responsibility of the EU member states. Nevertheless, it has a large, mature market in which market dominance in these two areas yields not only profit but also the promise of influencing foreign and security policies of the EU and its member states. Therefore, after a brief overview of historical relations between China and the EU, we summarize the key elements of the EU-China debates and the core issues of cooperation with China, classifying the EU-China debates as political or economic. We also examine the general business and legal environment faced by Chinese trading and investment firms at Single Market. Particular attention is paid to European grievances regarding access to the Chinese market and the newly agreed EU-China Comprehensive Agreement on Investment (CAI). In addition to the political and security concerns raised about technological cooperation with China, we approach and disentangle these complex issues by examining the EU's competitive position in technological development and the European regulatory framework for foreign direct investment.

The chapter titled "Economic and political interests of the major powers: the United States, Germany, and Russia" analyzes the main political and economic interests of the United States, Germany, and Russia with regard to technological confrontation/cooperation with China; this chapter also uses the terms geopolitics and geoeconomics. The original term "geopolitics" has become increasingly popular in recent years and refers to the politics or international relations that are influenced by geographical factors. After the collapse of the bipolar world in 1990, the rise in popularity was often explained as a return to normality, when balance in international relations rather than ideology was the key factors in understanding the main trends in global politics. At the same time, the absence of major wars reminds us that "history doesn't repeat itself, but it often rhymes," to paraphrase Mark Twain. In our understanding, the recent tectonic shifts in the world order can only be justified if geographically influenced economic factors are also considered. Therefore, the analysis in this chapter focuses more on the economic interests of the three countries. We are aware that geoeconomic and geopolitical factors are often not easy to separate. The chapter places particular emphasis on the reshoring of manufacturing capacity and technological decoupling, and it provides an overview of trade and investment relations with China in each case.

In "Chinese Investment and 5G cooperation in the EU: France, Germany, and Italy," we discuss the details of 5G policies in three countries, specifically Germany, France, and Italy, which we chose because these are the main EU economies. The main idea is that European dominance in 5G is sought by the United States and China because the size of the European market and dominance in this market could give them an advantage over their rivals in global competition. The EU, politically divided and economically advanced, is a key area to win the race for technological dominance. The EU does not have the opportunity or the tools to significantly influence this competition, and for this reason, this part focuses on market and policy analysis at the country level to make a proper assessment of the situation. Topics also include a general assessment of Chinese FDI in the three countries and a brief assessment of what these countries can offer to Chinese firms and what is the main motivation for Chinese investment in these countries.

Unlike previous parts, the chapter on "Chinese Investment and 5G networks in the Visegrád Countries"—at least in the introduction—delves much deeper into the history of the two parts of the European continent in order to show the differences between them and to explain the different economic development needs of the countries in the region. The introduction to this chapter aims to highlight the common elements in the historical development, which, in our understanding, lead to a different perception of China and ultimately to a slightly different legal and economic environment for 5G networks. The particular historical development of this region, as we understand it, also gives us a reason for why some countries seem to be reluctant partners in the EU, with special attention paid to the Visegrád Four. Subsequently, this chapter also discusses the diversification efforts of these countries, all of which are trying to expand their trade relations and investment flows to reduce their asymmetric dependence on the Western European countries.

The final chapter, "Conclusions," summarizes the findings of the previous parts and focuses on the geopolitical or geoeconomic conclusions. The main reason why the conclusions of the chapters are discussed here rather than separately at the end of each chapter is to show the complexity of the issues discussed in the book and how deeply these issues are intertwined. Particular attention is paid to the European implications of the competition for technological dominance. The first and last subsections here are not directly related to the previous chapters of the book; the first one (6.1. "Technology in the Chinese economic model") tries to contextualize the technological race between China and the United States, what is the relevance of this competition for China's economic model, and how new plans, the 14th Five-Year Plan and "Vision 2035", relate to technological development. The last subsection tries to formulate the challenges ahead and prescribe policy recommendations for the EU and its member states.

An interpretation of China's interests in technological competition are also embedded in the last chapter, the main message here being that the more the West tries to contain China in its economic and technological rise, the more difficult it will be for voices in China to argue for cooperation and peaceful coexistence with the West. Moreover, it will also be difficult to restart the globalization process, which has apparently stalled in recent years, without China.

The technological competition between the two superpowers also challenges the EU on many levels and raises questions that cut to the core of the EU's existence: What do we mean by the term "technological sovereignty" of the EU and member states; how can we achieve this and promote European flagship tech firms in the Single Market; how can the EU speak with one voice on China issues when member states' interests are so different; and how can member states' sovereignty in technology and economic development be reconciled with the drive to make the EU less dependent on other powers. However, it seems clear that the EU and its member states are much more dependent on the United States for technology than they are on China, which raises the question of why we get the opposite impression when reading news portals and foreign policy analyses. The main goal of this book is to answer this fundamental question, revealing in our understanding why the tools of geopolitics and geoeconomics are key to understanding the future.

Acknowledgments

The completion of this book would not have been possible without the professional help and support of my colleague Dean Karalekas and my daughter Nóra Moldicz, who happen to be my very best friends.

I would also like to thank Mikhail Karpov and Tamás Novák, who encouraged me throughout the project and whose professional advice I was able to seek.

I would also like to express my gratitude to Theodora Wiesenmayer for her encouraging contributions to the making of the book and her willingness to prepare the book's index.

My sincere thanks go to the management of Budapest Business School, University of Applied Sciences, especially to the Rector of the University Balázs Heidrich and the Head of Department of International Relations Balázs Ferkelt.

Csaba Moldicz

1   The battle for technology

The global arena[1]

"Who rules East Europe commands the Heartland; who rules the Heartland commands the World-Island; who rules the World-Island commands the world," Mackinder wrote at the beginning of the 20th century (Mackinder, 1919: 150). In today's world, an exclusive focus on geography cannot be justified, but the original sentiment, reflecting new realities, can be rephrased: "Who is at the forefront of technology rules the world economy who rules the world economy rules world politics, and who commands both economy and politics rules the world." Every facet of our life hinges on technology that we use and have access to. In the military, in business, in education, in transportation, and in health care—as the recent coronavirus pandemic painfully reminded us—technology surrounds us, and because we are reliant on these aspects of our lives, we are a technology-dependent species. No wonder that access to the latest and best technology is high on the agenda of every government, nation, and company.

Technological competition between the United States and China is, as we understand it, the most important field of competition for hegemonic power. This struggle has often been portrayed as a race between two different political regimes (Mead, 2014; Sun, 2019). Another aspect is that in this contest, although it plays out on many levels and in several places, Europe is wealthy and inherently weak as a political power, making it the ideal battleground for this final contest. For both the United States and China, Europe is a major business and investment partner, and therefore the European continent, especially the EU, is an idealistic spot for this conflict since the reward for the winner here is the highest on the globe. This chapter discusses the main narratives of technical and broader competition by drawing on the existing literature (see Section 1.1), then it investigates the "hard factors" of the competition by looking at key statistical indicators (see Section 1.2).

1.1 Technology as a key factor of a superpower's influence

The debates that shape and change our perception of China have intensified in recent years. With the Chinese reforms of 1978 and later, a new economic paradigm was born that brought immense changes and lifted hundreds of

DOI: 10.4324/9781003128625-1

millions of people out of poverty. Over the last four decades (1980–2019), the average GDP growth of the Chinese economy has been 9.44 percent. It is no wonder that the analysis of this economic miracle puts Chinese economic development at the center of research and raises questions such as: what are the reasons for such rapid growth; what are the economic and political factors; can it be emulated by other countries; and most importantly, where will this path lead? When will China's rise stop? Should it be stopped at all? How should the West, especially the United States, respond to the challenges China presents? Is the new Chinese economic development model superior to liberal free market systems? Section 1.1 will discuss three main narratives: the "rise of China" narrative, the United States' dilemma expressed in the "engagement" vs. "disengagement" debate, and the "developmental state" narrative. We will see that all three narratives add new aspects to the debate on understanding China's development, but with regard to the main research topic of this book, the developmental state debate promises to give us more insight into the nature of China's political and economic regime and thus predict the outcome of the contest between the two powers.

1.1.1 *"The rise of China" debate*

Different narratives of China's rise are part of an old debate, which has never abated and probably never will. The nature of this discourse is more academic, and its contribution to understanding China's technological rise is that it suggests a broad conceptual framework for us to comprehend the stakes and domains of this competition, but it offers no rationale for predicting which nation is most likely to emerge as the winner in the contest for technological supremacy. In this debate, scholars tend to focus on indicators such as GDP, GDP per capita, the role of the national currency in foreign reserves and international trade, various military indicators (size of the army, expenditures, technology deployed), and the ability to lead in areas of technological competition. This book focuses essentially on the last area, and also in this area on how recent developments of competition are played out in Europe and what is its significance in international relations.

From the early 1990s on, the debate over China's rise has polarized researchers, political leaders, and diplomats. The now-powerful country's economic, technological, and military advances have attracted interest beginning in the early 1990s. Nicholas Kristof put it enthusiastically as early as 1993:

> The rise of China, if it continues, may be the most important trend in the world for the next century. When historians one hundred years hence write about our time, they may well conclude that the most significant development was the emergence of a vigorous market economy—and army—in the most populous country of the world. This is particularly

likely if many of the globe's leading historians and pundits a century from now do not have names like Smith but rather ones like Wu.

(Kristof, 1993)

While Nicolas Kristof heralded the inevitable rise of China without hesitation, another team of researchers expressed serious doubts as to whether China can lead, especially in innovation. They pointed out that the political environment in China imposes severe constraints on schools, universities, and businesses (Abrami et al., 2014). They were not alone in reasoning this way. To mention just one of the most influential China experts of the time, Elizabeth Economy emphasized the problems created by heavy state intervention in the economic system, pointing to the Chinese state's hunger for capital, which they believed was starving the more efficient private sector (Economy, 2019). A similar view is expressed by Julian Baird Gewirtz, who reminds us of the challenges posed by:

> top-down CCP-led innovation … waste and massive oversupply, or the challenges of supporting small entrepreneurs and researchers without heavy-handed interference.

(Gewirtz, 2019)

This is the point at which the debate becomes linked to the question of the Chinese (economic and political) model. The main argument of those prophesying the collapse of China is that only open, democratic systems can sustain economic institutions and guarantee long-term growth (this concept is discussed in more detail in the chapter on the developmental state debate).

The following thread in the debate over China's rise is whether the country can rise peacefully, or whether the containment of China is unavoidable due to Beijing's behavior. The discussion also revolves around whether China poses a threat to world peace and order. Realists tend to argue that the country's rise will lead to hegemonic war and that a rising China will ultimately pose a menace to world peace, while libertarians envision a more peaceful nature of the globalized world in which states seek compromise and are willing to cooperate (Ye, 2002).

This debate has resurfaced in various forms over time, including even to extremes.[2] The question remains as to how much we can gain from such a broad debate, encompassing so many voices and viewpoints. As we understand it, such a debate is futile, as no right answers can be given, but it draws our attention to what the areas of contest are between the United States and China (economy, military, and technology) and the fact that China's rise cannot continue without the help of technology development, the most important factor in Chinese economic development. This may be precisely the conclusion of American policymakers, who seem to prefer containment and decoupling from China to a policy of engagement.

1.1.2 The dilemma of "engagement versus decoupling"

The political debate of "engagement versus decoupling" policy offers us a different angle on the discourse, one that is rather US-centered. Assuming the inevitable rise of China, participants in the debate are mostly trying to give answers on how to deal with the consequences of that rise. In this debate, libertarians usually tend to argue that the policy of engagement is the best strategy for dealing with the threat posed by the inevitable rise of China. This was not the approach taken by the US administration of President Donald Trump between 2016 and 2020, which put the United States on a collision course with China and sought or forced disengagement from China. The intensity of the policy dilemma reached new heights after 2017 when the Trump administration revised the US National Security Strategy, and the next peak came during the Covid-19 pandemic, when Secretary of State Mike Pompeo in his speech "Communist China and the Free World's Future" in July 2020 took aim at engagement policy:

> We must admit a hard truth that should guide us in the years and decades to come, that if we want to have a free 21st century, and not the Chinese century of which Xi Jinping dreams, the old paradigm of blind engagement with China simply won't get it done. We must not continue it and we must not return to it.
>
> (Pompeo, June 25, 2020)

The roots of engagement policy can be found in the Nixon era, but it began to take on a sharp profile in the 1990s when the Soviet Union collapsed. The end of the bipolar world led to the reconceptualization of US policy toward China. George H.W. Bush introduced the term "engagement" with respect to the People's Republic of China (PRC), but it was not until the Clinton Administration (1993–2001) that the term "engagement policy" became associated with the idea of economic changes and reforms that were followed by political changes, democratization in China (Neil, 2019). Finding a place for China in the post-1990s world order was the best way to usher in China's democratization, and so thus a unique new policy was introduced and a unique new word to describe it "congagement": a portmanteau of "containment" and "engagement." To be fair, even during the Clinton administration the idea of engagement was controversial. The 931-page Cox Report described China's threats in detail in 1999, and despite its articulated concerns, the Clinton administration remained committed to engagement (Select Committee, US House of Representatives, 1999).

Moreover, democratization was described as a natural side effect of long-term economic success; otherwise, economists argued, an economic rise would simply cease. This line of reasoning was first used by Lipset, who linked economic success to the emergence and maintenance of democratic, pluralistic institutions (Lipset, 1959). But today, there are newer versions of this idea (Ferguson, 2011; Acemoglu & Robinson, 2012). The concept remains the same: there can be no long-term success without the account-ability of the political elites, as the lack of it leads to the establishment of exclusive economic institutions giving short shrift to innovation and new ideas. The debate over the adequacy of the policy was never settled, but a dramatic shift in US foreign policy toward China came when the aforementioned National Security Strategy redefined the perception of China's role in 2017. The latest (2020) version of the strategy mentions China twice. On page 1, the strategy links this struggle between the two powers to the issue of technology:

> A more powerful and emboldened China is increasingly asserting it-self by stealing our technology and intellectual property in an effort to erode United States economic and military superiority.
> (National Counterintelligence and Security Center, 2020: 1)

And on page 2, the strategy underlines that China, along with Russia, is using all instruments of power (including intelligence capabilities) to target the United States. Since then, the alleged failure of the engagement policy has often been pointed out. Thomas Neil maintains that the enforcement of American interests was of greater concern to Bill Clinton, with whom we closely associate engagement policy (Neil, 2019). He also reminds us that the United States was able to convince China to cooperate on many issues: law enforcement, disease control, environmental issues, and military maritime security, just to mention a few.[3] In other words, the engagement policy did not create democracy in China, but it did bring about significant economic benefits to the United States, while economic policies in the United States failed to compensate the losers of globalization at the expense of its winners.

As the rise of China did not stop, the tone of the debate changed, and a bipartisanship seems to have emerged on the China issue. The new Biden administration will most likely retain certain elements of Trump's policies, and containment in areas of strategic importance (5G, artificial intelligence [AI]) will be retained as a crucial tool of China policy.

1.1.3 The "developmental state" narrative

The debate about the "developmental state" has been forgotten in academia because it has been argued that globalization has rendered states weak and powerless, and thus the days of the intervening state are over. Yet today, after the Covid-19 pandemic, states appear stronger than ever. Interventions in every segment of the economy have been shown to be possible, and economic activities are being regulated to a degree that no one thought possible. Questions have been raised about who can and cannot provide technology-intensive services. Instead of the attraction of foreign direct investment (FDI), the rules of FDI screening are being debated. Bringing home global supply chains is fiercely debated. All this requires clear and strong interventions, and these interventions are also part of the economic development strategy.

The third China narrative broadens the picture and brings us closer to the core content. In this understanding, the real question is what China is as a country in the modern world, how we define its political and economic structures, and whether we can find similar models in the world. The two focal points are the changing role and power of the state in a globalized world and the question of a model or parallels in the world that might give us clues about what to expect in the later stages of China's development.

It has often been argued that the main reason for cooperation with China is the changing nature of the global order, in which globalization, the argument goes, would have changed the attitudes of the main actors, the states. Jiang Ye, a professor at the Humanity College at China's Shanghai Normal University, claims that China seems more willing to cooperate than the established powers:

> But for all the strength of states as the principal actors in the international system, the dominance of states as the focus of political authority is declining with the impact of globalization on the international system since the end of the Cold War.
>
> (Ye, 2002)

As a counterpoint to this argument, the developmental state model is about the emergence and success of the strong state that is willing to intervene efficiently in economic and social development. In our understanding, China can be interpreted as a uniquely altered variant of the Asian developmental state model. Looking at the traditionally accepted characteristics of so-called developing states,[4] the following distinguishing and common features can be identified when comparing the Chinese economic model with the original developmental states of Asia:

strong economic planning tools. Economic planning tools are stronger in China than in the original development states/economies (i.e., Japan, South Korea, Taiwan). Not only do state-owned enterprises have a larger role in the economy,[5] but the Chinese government still has a double-track, price-setting system[6] that distorts market prices while at the same time it helps guide such enterprises into new sectors. Similar systems existed in the 1950s and 1960s in Japan, South Korea, and Taiwan. Kasahara underlined that these states were involved in some form of price fixing, strong regulation of labor, and foreign exchange and financial markets (Kasahara, 2013: 1). A special and often noted tool is a devaluation policy that ensures more competitiveness to the domestic companies (Johnson, 1999: 32–60). In the case of China, too, the devaluation policy was featured by the United States as currency manipulation (Setzer, 2019).

1   *Cheap labor.* Although academic papers have already pointed out that wages and salaries are rising, the Chinese economy is still dominated by cheap labor. Despite growing wages and salaries, there is still a backlog of cheap labor in China's rural sector. According to the World Bank, agricultural employment accounted for 24.73 percent of total employment in 2020. In advanced countries, this figure is usually less than 5 percent. In addition, the relevance of this factor in competition will decrease as more companies will use robots in manufacturing.

2   *Export orientation.* It is strong in each developmental state, although the Chinese market is less open than other Asian markets, which is understandable given China's size and different historical development path. Learning from the bad experiences of the 1930s and before, China has always been wary of opening its economy too quickly to foreign capital, while establishing the first special economic zones in Guangdong and Fujian and later expanding this model to other areas of China.

3   *Land reform* as the starting point of economic reform and take-off. Land reform was crucial in each case, as Japan, Taiwan, and South Korea began their respective industrialization processes with comprehensive land reforms completed in the 1950s and 1960s. Although the Chinese have taken steps to modernize the agricultural sector in recent years, the process is far from complete.

4   The question of *bureaucracy and the rule of law.* The connection between democratic institutions and growth rate is not present/absent in the case of China, but this connection is not a critical component of the developmental state model. The rule of law and the relative independence of the state bureaucracy, as well as meritocratic, merit-based selection, are more important inherent elements of the developmental state model. However, China's performance in these areas is weak, with corruption in particular currently overly prominent.

It can be argued that the Chinese economic model is unique due to the size[7] of the country and its historical development, although it bears a strong resemblance to the original developmental state model followed by other advanced Asian economies. The model can be used efficiently as a lens through which to view the Chinese economy, and the resemblance is even more striking when one considers how much the global economy has changed over the decades. Therefore, in our understanding, the Chinese economy can be viewed as a special case of the developmental state in the 21st century. The differences between China and the three original models of the developmental state would not be outstanding if one did not consider the freedom of maneuvering room for economic policy which comes from the size of the economy.

Why is the developmental state model important in terms of technology competition between the United States and China? It provides us with a model in which technology development along with strong state intervention are extremely important explanatory factors. The developmental state model can also help us understand why the absence of democratic institutions may not necessarily impede China's technological and economic development, but rather has the potential to fast-track it. The consensus around the original developmental state model—built on the experiences of Japan, Korea, and Taiwan—also draws our attention to the importance of a farsighted and quasi-autonomous bureaucracy that can formulate the right incentives for research and innovation. In our view, the emphasis should be on the rule of law, particularly laws that are enforceable in the economic sphere rather than Westminster-style democratic institutions.[8] This could also help us to rid ourselves of biased assessments based on ideology reinforcing the idea that communist ideology plays a significant role in Chinese economic development.

Having reviewed the main points of competition between China and the United States and recognizing that China's behavior is best understood in terms of the "developmental state paradigm," let us next examine why Europe, especially EU, has become the main battleground in this struggle.

1.1.4   The EU, the excellent "prey"

Not only the EU, but the larger European region—including the United Kingdom, the European Free Trade Association countries, and the West Balkans, whose countries have not yet joined the EU—form a more or less coherent region in which ethnic, linguistic, and religious diversity are the predominant features of the region. In this region, we do not find a single large country capable of assuming a global leadership role in technology development. In the EU, Germany is the only country that could hold its own in this competition to a certain degree; however, Germany does not have the necessary resources and especially the economic size to formulate this ambitious goal. The EU would obviously have this optimal size of domestic market for the technology-intensive products and services of European firms, as well as the financial and human resources to participate in the race for technological superiority, but institutional problems and policy failures have limited this European effort since the early emergence of European integration.

The main reason for this policy failure is that the EU does not have a full-fledged industrial policy and therefore cannot efficiently intervene in the economy or guide economic actors as the United States or China can on their own if necessary. The lack of an industrial policy is often blamed on weak cooperation efforts of EU members; however, this argument is misleading. The Single Market was originally created based on the idea that equal opportunities must be created for every firm and company, regardless of their national origin. This is the reason why EU competition policy has two pillars: one that regulates firms (mergers, acquisitions, etc.) and the other that sets the rules for government intervention in the economy. Each EU member country holds more or less the same maneuvering room in state intervention, which makes sense because without this part of competition policy, stronger, wealthier countries could reshape (and distort) the market to their own advantage with impunity. Soon enough, the Single Market would be dismantled and the EU would cease to exist.

In other words, the weakness of European industrial policy is embedded in the institutions of the EU, and as long as the EU exists, none of its countries can truly aspire to global technological leadership due to nonexistent industrial policy. None of the countries would be able to become a global leader in the case of a fragmented Single Market due to the lack of optimal market size. This is a catch-22 situation. There is an additional problem: in sectors where new technologies are crucial, the EU or EU countries have failed to create globally successful EU firms over the past two decades. Maincent and Navarro put it this way in 2006:

> the EU does not suffer from the absence of large world-class companies, but rather from the absence of growing companies in new high-technology industries. With a few exceptions, the EU has not been able to promote the emergence of international players in the fast-growing sectors of the economy.
>
> (Maincent & Navarro, 2006)

Weiss argues that with the new European Commission from 2019 on, there is theoretically a chance that a more direct approach to industrial policy can be formulated, adopted, and implemented by EU member states, especially as the Brexit has created space for the implementation of the Franco-German model, in which the creation of "national champions" has been a predominant key feature (Weiss, 2019). This change of mindset is clearly evident in the new economic strategy of the German government published in early 2019. The document places a very strong emphasis on the social market economy (also called Rhine capitalism), based on the idea of combining free market capitalism with social measures. The strategy stresses that national and European champions, while important, have not been made in recent decades; even the success story of Airbus is now very old, the document says. In conclusion, the German government underlines the need to revise EU state aid rules to create larger European companies, while tightening rules on domestic takeovers to protect strategic industries (Federal Ministry for Economic Affairs and Energy, 2019). At the same time, we are skeptical that this policy change can be enforced at the EU level without disrupting the Single Market or further limiting member state sovereignty.

Protecting European industry and relaxing state aid rules—which are necessary to create national champions—seem like a reasonable policy in the aftermath of Covid-19, but the question remains how long and profound distortions in the Single Market can be tolerated before it leads to growing tensions between member states for two reasons:

- As small nations with smaller budgets are much more constrained in the creation of their national champions, and diminishing economic advantages make economic integration less and less attractive to these countries.
- Smaller and less developed economies, Central Europe, have always relied on capital and technology imports from Western Europe, so restricting FDI in their case does not seem to be a reasonable policy. Moreover, it is against their fundamental interests to diversify trade and investment relations with other parts of the world economy. We must bear in mind that the differences in economic interests between Western and Central Europe are obvious and they are unlikely to disappear in the near future.

So far, it can be shown that the EU is divided in many ways, and the institutional design of the Single Market hinders the implementation of an efficient industrial policy which is unlikely to change anytime. At the same time, the Single Market is still the second largest market in the world too.[9] Therefore, any multinational company that wants to be a world leader in its segment must have an excellent position in the EU market. So, it can be argued that there are four elements combined that attract China and the United States and make the EU an excellent "prey":

- the political and military weakness of the region;
- the lack of a coherent industrial policy;
- the different economic interests of the member states; and
- the wealth of the region.

In the following subchapters, we investigate how well prepared the United States and China are to capture their "prey." In doing so, we not only analyze basic technology-related figures, but also discuss the main contours of their historical technology developments, economic development strategies, and their motivations in Europe.

1.2 Historical and recent technology developments in China and the United States

1.2.1 China

China's hunger for technology has been frequently highlighted by researchers and analysts over the past four decades. Modern China's eagerness to absorb foreign technology and accelerate its domestic technology development is a logical step in its economic development. The need for rapid development and the recent US decoupling strategy justify an import substitution strategy in technology as well.[10]

The main dilemma in the economic development of modern China has been how to accelerate technological development without creating asymmetric dependence and increasing reliance on Western technology and capital investment. This problem can be easily captured in the dilemma of "Treaty Port" vs. "Manchurian" industrialization (Naughton, 2007: 43). The so-called treaty ports of the 19th and early 20th centuries—enclaves of the Chinese economy under foreign rule—while economically successful, severely limited economic sovereignty and led to lopsided development that created an asymmetric dependence of the Chinese economy. These enclaves specialized in consumer goods, while "Manchurian-style industrialization" emphasized heavy industry and heavy state intervention. "Manchurian-style development" began in the Japanese puppet state of Manchukuo, but after the Civil War (1949), the Chinese Communist Party (CCP) overtook and "improved" the practices of the Chinese Nationalist Party, the Zhongguo Guomindang (KMT). These economic development practices emphasized heavy industry; state-owned enterprises and state guidance were favored over the private sector and market forces. Until the early 1960s, heavy industry development (associated with the Big Push theory[11]) was assisted by significant economic and technical aid from the Soviet Union, but this assistance was abruptly ended by growing tensions between the Soviet Union and China and disagreements over the future of communism in 1960 (Pantsov & Levine, 2012: 536).

When Xi Jinping, the President of the PRC, first called for self-reliance in 2014, the concept of self-reliance already had a long history (Laskai, 2019, June 19), going back at least to Mao.[12] In retrospect, we can understand that this is not the first time China has been denied cooperation in the field of technology:

- For example, the Soviet Union stopped providing technical assistance to the PRC during the Cold War, but China was able to continue developing nuclear weapons.
- The United States pursued the same strategy in 1991 when China wanted access to satellite and rocket launch technology. The United States imposed comprehensive sanctions on the export of satellite components to China (Fitzwater, 1991: 446), but these were not sufficient.[13]

In the long period of economic reform from the late 1970s to the early 2000s, the economic development strategy slowly began to lose its socialist, heavily interventionist character, yet it never became a laissez-faire economy in which market forces are the ultimate forces of economic development. Elizabeth Economy summarizes the incompleteness of China's economic reforms in 2018:

Economic reforms have progressed in some areas, notably toward a market-driven currency, greater market openness in areas such as hospitality, increased access to the Chinese stock and bond markets for foreigners, and more discretion for banks to set interest rates based on the creditworthiness of the borrowers. Yet in the majority of these areas, there has also been backsliding and the reintroduction of some state controls.

(Economy, 2018: 152–153)

In some cases, the critical comments contradict each other, which is the case when it comes to the "heavy-handedness" of Chinese technological development. For example, Julian Baird Gewirtz points out the problems of top-down, CCP-driven technological innovation,[14] while noting that China could quickly move up the value chain:

But China has quickly moved up the value chain, creating world-class industries in everything from 5G and artificial intelligence to biotechnology and quantum computing. Some experts now believe that China could unseat the United States as the world's leading technological force. And many U.S. policymakers view that prospect as an existential threat to U.S. economic and military power.

(Gewirtz, 2019)

The same planning attitude or "heavy-handedness" in technology development is evident in the so-called "Made in China 2025" strategy. Although the plan was inspired by the German Industry 4.0 plan (adopted in 2013), it differs from the German version and also from China's earlier technology development strategy (Kennedy, 2015).[15] On the one hand, the "Made in China 2025" strategy is broader than its German counterpart, as the competitiveness of Chinese firms is more uneven, and it also has to deal with challenges from Asian "low-cost" producers. On the other hand, the plan also sets a target to increase the domestic content of core parts and other materials to 40 percent by 2020 and 70 percent by 2025. It must also be mentioned that the plan targets ten priority sectors and the information technology sector is one of these key sectors.

The Chinese strategy is very clear in its goals, and the strong state guidance brings this strategy very close to the original French or Japanese economic development strategy of the 1960s and 1970s. As we mentioned earlier, the original developmental state model bears a strong resemblance to what China is now trying to achieve and implement in a globalized world. The irony is that American decoupling policies are now more than ever creating conditions favorable to the implementation of China's state-centered, self-reliance-oriented economic strategy, and those within China who would push for more reform and opening may find themselves in a situation where it is impossible to reasonably argue for granting more market access to Western European or American firms in China.

We argued earlier that the EU suffers from the lack of an efficient industrial policy due to institutional constraints, in the case of China, we argue, China suffers from a heavy-handed industrial policy which combined with the external pressure from the United States (decoupling) can reinforce the instincts of Chinese politicians not to open the economy further, but rely more on homegrown technologies.

1.2.2 The United States

Not only has the Chinese approach to economic development, and especially to technology development policy, changed dramatically, but the course of American technology development has changed several times in recent decades. The rise of neoliberalism in the 1980s and the collapse of the Soviet Union in 1991 challenged the role of government in promoting economic development. The neoliberal renaissance meant a shift from manufacturing to knowledge-based services, increasing economic linkages between the American and Chinese economies.[16] The bursting of the dot-com bubble did not change the course of policy in the early 2000s; the Global Financial Crisis was needed to significantly alter the American approach to economic development. In the aftermath of the Global Financial Crisis, there was considerable debate about whether the response to the crisis could be interpreted as a revival of Keynesianism: since state intervention was more limited, it relied more on monetary policy measures—quantitative easing—than on fiscal spending. For this reason, Bello is critical of Obama's half-hearted Keynesianism, calling it "technocratic Keynesianism" that:

> resulted in a protracted recovery, continuing high unemployment, millions of foreclosed or bankrupt households fending for themselves, and more scandals in a Wall Street where nothing had changed. Obama did not pay for this tragic outcome in 2012, but Hillary Clinton did in 2016.
> (Bello, 2017: 31)

During the Obama administration, some policies were adopted with a more direct approach to technology development. The National Innovation Strategy was first adopted in 2009 and updated in 2011 and 2015, but since then the Trump administration has not announced a formal innovation strategy. In September 2019, the Council on Foreign Relations released its report "Innovation and National Security: Keeping Our Edge." This asserts that there is a risk of falling behind China in technology development. The report provides the reader with a historical perspective as it points to the decline in federal spending on research and development (R&D):

> … as a percentage of GDP peaked at above 2 percent in the 1970s and has declined since, from a little over 1 percent in 2001 to 0.7 percent in 2018. In 2015, for the first time since World War II, the federal government provided less than half of all funding for basic research.
> (Council on Foreign Relations, 2019: VI)

Table 1.1 R&D expenditure in China and the United
States (in percentage of GDP)

Country name	1998	2008	2018
China	0.65	1.45	2.19
United States	2.50	2.77	2.84

Source: Own compilation based on World Bank database.

Table 1.2 Gross domestic spending on R&D in China and the
United States (million US dollars)

Country name	2000	2018	Change in percentage (%)
China	44,442	526,063	1,083
United States	361,469	551,518	53

Source: Own compilation based on OECD database.

It should be noted that the composition of R&D spending is important because it is not the relative or absolute level of funding that is important but rather the public share that has declined in recent decades. Table 1.1 illustrates the R&D expenditures of the two countries. According to the World Bank, China's R&D spending was 2.19 percent of GDP in 2018, compared with 2.84 percent in the United States in the same year.[17]

In the case of China, it is not only the amount spent on R&D that is breathtaking, but also how fast that amount has grown over the past two decades. The R&D expenditures were only 0.65 percent of GDP in 1998 and 1.45 percent in 2008, while the change in the United States over the same period was much smaller.

The American lead remains significant, as reflected in the absolute numbers of R&D expenditures, too. In 2018, the United States spent $551.5 billion on R&D, while China's spending was $526 billion, according to Organization for Economic Cooperation and Development (OECD) data. The growth in nominal R&D spending has again been tremendous, as China spent only $44,442 million on R&D in 2000, which basically means that the country's R&D spending has expenditures multiplied by a factor of almost 11. Table 1.2 shows the development of gross domestic expenditure on R&D in US dollars.

The analysis concludes that China will be the world's largest spender on R&D, adding:

China is closing the technological gap with the United States, and though it may not match U.S. capabilities across the board, it will soon be one of the leading powers in technologies such as artificial intelligence (AI), robotics, energy storage, fifth-generation cellular networks (5G), quantum information systems, and possibly biotechnology.

(Council on Foreign Relations, 2019: 5)

Two crucial differences in structure can be discerned. (1) The structure of spending is different. The United States spends much more on basic research, 17 percent, while basic research accounted for only 5.5 percent of China's R&D spending in 2017. China spent 84 percent of its funds on experimental research in the same year, while this share was only 63 percent in the United States in 2017 (National Science Board, 2020: 30). (2) In the case of China, private enterprises are more dominant. The business sector is the dominant source of spending, with 76 percent of R&D spending coming from the business sector in 2017, compared to 62 percent in the United States (National Science Board, 2020: 32).

The two differences can be linked, as less direct control over R&D spending in China leads to a shift in focus to R&D that is faster for private firms to monetize. It should also be noted that business R&D spending in China is heavily concentrated in manufacturing (87 percent), while in the United States only 67 percent of business R&D spending was in this sector, based on 2016 data from the National Science Board.

1.2.3 Human resources

Besides the amounts of direct spending on R&D, there are other—rather soft—figures that help us understand the development and research potential of the two countries. The following list contains the most relevant human resources-related indicators for the technological race between the United States and China:

1   *Science & Engineering degrees.* According to the Science & Engineering (S&E) Indicators of the National Science Board, 22 percent of S&E bachelor's degrees were earned in China, placing China second in the global ranking in 2015 (India: 25 percent, EU: 12 percent, and the United States: 10 percent). The difference is not surprising, given the size of the population, but what is more remarkable is that about half of all bachelor's degrees in China are earned in S&E, compared to only one-third of degrees in the United States (National Science Board, 2018: 8).

2   *S&E doctoral degrees.* According to the 2016 estimates, 1.7 million S&E degrees were awarded in China, compared to about 800,000 in the United States and one million in the EU. As we can see, China is heavily focused on manufacturing R&D, so it's no surprise that nearly 70 percent of S&E degrees come from manufacturing. China is the third largest producer of S&E doctoral degrees in the world, the EU leads with 70,000 doctoral degrees, while the United States is second with

40,000 doctorates, according to the 2016 data. China produced 34,000 doctoral degrees in 2015; however, China now leads in S&E doctoral degrees (National Science Board, 2020: 4–5).

3   *Publications.* As with the trends in science and technology degrees and R&D expenditures, publication output (per peer-reviewed article) has grown rapidly in China. Since 2000, output has increased almost tenfold, and China has overtaken the United States. According to the US National Science Foundation (NSF) reports, China produced the most scientific publications in 2017, but the lead was already taken by China in 2016, when about 426,000 papers were published by Chinese scientists, compared to 409,000 in the United States in 2016. China's specialization in engineering can be seen in its publication output. China produces more than twice as many engineering-related articles as the United States, while the United States surpasses China in biomedical and health sciences. In addition to quantity, quality also plays a role here: the global share of the top 1 percent articles is often measured as a share of the global share. If the number is 1, it means that the two shares are equal. In the United States, the index was 1.9 in 2016, while in China it was only 1.1 in the same year, but we must keep in mind that this index was 0.4 in 2000 (National Science Board, 2010: 12).

4   *Patents.* China has been filing the most patents in the world for many years. Given the size of the country, it should come as no surprise, but 2019 was the first year China filed the most patents under the Patent Cooperation Treaty (PCT) and also overtook the United States. The relevance of PCT patents stems from the fact that a PCT patent can be seen as an indication of a desire to expand into new markets. According to the World Intellectual Property Organization (WIPO), China specializes in digital communications, computer technology, and audiovisual technology (WIPO, 2020).

We could see that the Chinese advantage is not only based on low labor costs, but knowledge and innovation are also reasons why the country could become a manufacturing hub. The Trump administration's recent push to rip control of the global supply and manufacturing chains from China could face serious challenges, as American companies doing business in China have very clear advantages there: a large market, relatively low labor costs, clusters of firms that allow seamless collaboration, and the spread of information among firms and inventions needed to implement new technologies or other forms of innovation. Willy C. Shih, a professor of management practice at Harvard Business School, summarizes the complexity of the American effort as follows:

A consequence of these complex interdependencies is a deep tiering of supply chains, with manufacturers dependent on their first-tier suppliers, which, in turn, are dependent on a second tier, which are themselves dependent on a third tier, and so on. Visibility into third, fourth, and more distant tiers is challenging, making wholesale replacement of anyone in the chain, let alone the entire chain, extremely difficult.

(Shih, 2020)

Even if the American authorities find a way to give the necessary incentives (tax incentives, possible re-shoring subsidizes) for companies to leave China, it will take from cc. 3–8 years before we see the results of this policy. We also need to keep in mind the magnitude of the US effort, as about a third of the world's manufacturing output now comes from China.[18]

1.2.4 R&D at the corporate level

1.2.4.1 Market value

The Forbes Global 2000 list ranks publicly traded companies according to their global importance. To do so, it relies on four metrics: sales, profits, assets, and market value. While there are only four American companies in the top ten most valuable companies in the world, there are five Chinese companies listed in this group. However, these are all in the banking or insurance industry. This means that these companies are in the top league because of the size of the Chinese economy, not because the particular company would be a leader in either internationalization or R&D. (There are 315 Chinese companies in the 2,000 most valuable companies, while 545 American companies are in this group.) (Forbes, 2020)

Table 1.3 Most valuable tech companies based on market value in 2020

		Market value in US$ billions	Headquartered in
1	Apple	2 trillion	US
2	Amazon.com	1.6 trillion	US
3	Microsoft Corp.	1.6 trillion	US
4	Alphabet[a]	1.05 trillion	US
5	Facebook	760 billion	US
6	Alibaba Group Holding	730 billion	China
7	Tencent Holdings	650 billion	China
8	Taiwan Semiconductor Manufacturing Co.	393 billion	China[b]
9	Samsung Electronics	339 billion	South Korea
10	Nvidia Corp.	310 billion	US

Source: Divine, J. (2020). Data retrieved on September 17, 2020.
Remarks:
a Alphabet encapsulates Google.
b Although Beijing claims Taiwan as part of China, the assessment of China's technological development would be distorted by Taiwan Semiconductor Manufacturing Company's inclusion in China's flagship group.

Looking at the crème de la crème of companies that specialize in new technologies or are considered pioneers in their segment, one can understand from a different perspective which countries are good at shaping the contours of the future. This segment of tech companies emerged in the 1990s when technologies related to the Internet and mobile telecommunications were born and quickly became a part of our lives. Since most of these technologies originated from American firms, the segment was dominated by these firms for decades, and there are few Asian firms that grew rapidly after 2000 and joined the elite club of firms. Table 1.3 illustrates the market value of these firms. It can be observed that two Chinese firms, one Korean tech firm and one Taiwanese tech firm are among the top ten most valuable tech firms. This is a significant change compared to the pre-2000 period. Large Chinese firms have undergone a rapid and very successful internationalization process over the past two decades, but the majority of the most valuable tech firms still come from the United States and not China.

1.2.5 R&D expenditures of corporations

When assessing technology competition between the two countries, it is important to consider not only the market value of each firm, but also the amount spent on R&D. The 2020 EU Industrial R&D Investment Scoreboard contains a comprehensive database of 2,500 companies worldwide that invest significant resources in R&D. Table 1.4 shows the R&D ranking of the world's top 2,500 companies. In 2019, 536 Chinese firms and 775 American firms were on this list. Chinese firms spent $118 billion on R&D, while American firms spent almost three times as much ($348 billion) as the Chinese in this area. Only Huawei is on the list of the top 20 companies (European Commission, 2020).

1.2.5.1 Venture capital investments

One particular business sector is the small technology company segment with great need for fresh capital, but at the same time with great potential for future growth. The gap between the Chinese and US venture capital markets has narrowed over the past decade, data shows. According to data from the Global Private Equity Report 2019, Chinese firms made up just 4 percent in 2014, which widened to a staggering 32 percent in 2018. Due to this trend, the gap between the United States and China has been closing fast, although the United States is still number one investor in this segment at 47 percent (2018). In 2018, China's private equity sector was dominated by Internet and tech companies, with 83 percent of total venture capital invested in them, while the rest (17 percent) went to other sectors (Bain & Company, 2019).

Table 1.4 R&D expenditure ranking of the world top 2,500 companies

	Firms	Country	Sector	US$ billion
1.	Alphabet	US	Software & Computer Services	23,160
2.	Microsoft	US	Software & Computer Services	17,152
3.	Huawei Investment & Holding	China	Technology Hardware & Equipment	16,712
4.	Samsung Electronics	South Korea	Electronic & Electrical Equipment	15,525
5.	Apple	US	Technology Hardware & Equipment	14,436
6.	Volkswagen	Germany	Automobiles & Parts	14,306
7.	Facebook	US	Software & Computer Services	12,106
8.	Intel	US	Technology Hardware & Equipment	11,894
9.	Roche	Switzerland	Pharmaceuticals & Biotechnology	10,753
10.	Johnson & Johnson	US	Pharmaceuticals & Biotechnology	10,107
11.	Daimler	Germany	Automobiles & Parts	9,630
12.	Toyota Motor	Japan	Automobiles & Parts	9,058
13.	Merck Us	US	Pharmaceuticals & Biotechnology	8,235
14.	Novartis	Switzerland	Pharmaceuticals & Biotechnology	7,713
15.	Gilead Sciences	US	Pharmaceuticals & Biotechnology	7,394
16.	Pfizer	US	Pharmaceuticals & Biotechnology	7,372
17.	Honda Motor	Japan	Automobiles & Parts	6,835
18.	Ford Motor	US	Automobiles & Parts	6,587
19.	BMW	Germany	Automobiles & Parts	6,419
20.	Robert Bosch	Germany	Automobiles & Parts	6,229

Source: European Commission (2020). EU R&D Scoreboard. The 2020 EU Industrial R&D Investment Scoreboard.

1.2.5.2 5G and AI

In contrast to the previous segments, 5G appears to be dominated by two Chinese giants, Huawei Technologies Co., Ltd. and Zhongxing Telecommunication Equipment Corp. better known as ZTE. It is estimated that the combined market share of the two companies in the 5G infrastructure market was around 40 percent in 2019 (Benner, 2020). The Dell'Oro Group report focuses on the phone equipment market, where Huawei had a 28 percent share and ZTE Group had a 10 percent market share in 2019. This segment is dominated by five manufacturers (Huawei, Nokia, Ericsson, Cisco, and ZTE). The equipment market covers areas other than 5G, but each company's market share reflects its broader market position and is meaningful in terms of its position in the 5G market. TrendForce also estimates market shares for 5G base stations. The estimated shares are very close to the market position in the global telecom equipment market (Huawei: 28.5 percent, ZTE: 5 percent in 2020; cited by Davies, 2020).

Patent activity is also an important factor in competition in this field and a very good indicator of who will win the 5G race, or at least lead it for a while. Research firm GreyB and Amplified have published a report titled "Who Owns Core 5G Patents? A Detailed Analysis of 5G SEPs in 2020" The authors of the report conclude:

> Huawei is leading with the most declared 5G patents i.e., 2,386 patent families followed by LG and Samsung with 1,388 and 1,353 patent families respectively. Ericsson is almost equal to Samsung and secured the 4th position with 1,350 patent families, while Qualcomm and Nokia have 5th and 6th place.
>
> (GreyB, 2020)

Based on market trends, it would be a given to predict China's eventual dominance of the global market. However, it is clear that market gains can turn into losses if the political environment restricts market competition for political reasons.

Both the United States and China are spending significant sums on AI development. According to ABI Research, China spent $7.4 billion on AI in 2018, while the United States spent only $9.7 billion in the same year. The amount spent meant that 52.3 percent of the market share went to the United States that year. Recent data from ABI Research shows that state guidance of the market does not always produce the desired results (ABI Research, 2019, October 3).

When the Chinese State Council issued the so-called "New Generation Artificial Intelligence Development Plan" in 2017, which included an ambitious roadmap with goals through 2030, commentary targeted the potential political threats and assumed that China would dominate in this area as well, especially in 2017 when China led the world in AI investment. In the long run, the United States appears to be leading this race: Between 2015 and 2019, 56 percent of AI investment came from the United States, while China's global market share was 22 percent (the United Kingdom: 6 percent, France and Canada: 3.3 percent; Tech Nation data cited by Sawers, 2020).

In addition to the value of AI investment, other factors (human resources, adaptation, development, research facilities, etc.) can also be considered.

Table 1.5 Revenue share in global telecom equipment market (percentage)

	China	*United States*	*European Union*
Talent	3	1	2
Research	3	1	2
Development	3	1	2
Adaption	1	3	2
Data	1	2	3
Hardware	2	1	3

Source: Castro et al. (2019).

Danial Castro compared China, the EU, and the United States in this area (see results in Table 1.5) and their main finding was that:

> despite China's bold AI initiative, the United States still leads in absolute terms. China comes in second, and the European Union lags further behind. This order could change in coming years as China appears to be making more rapid progress than either the United States or the European Union.
>
> (Castro et al., 2019: 2)[19]

We have seen in this chapter that the three narratives of China offer us different angels of the debate on how to understand China. While the "China rising" debate helps us contextualize the race in technological development between China and the United States and shows us the main areas of competition, looking at the US foreign policy debate helps us understand that competition will not stop in the coming years regardless of the administration in office. The "developmental state" paradigm has proven useful in demonstrating why China is so focused on technological development as a key area of its economic development. If one can strip away the ideological layers of the debate, we can understand that the arguments of the American critique are not new, nor are they specific to China. Similar criticisms were made in the case of Japan and Germany in the 1960s, 1970s, and 1980s when they were ascending. The "developmental state" approach has shown us that there is no direct and necessary correlation between economic success and Western political institutions. China might be able to continue its rise without adapting a Western-type democracy. The analysis of the data also showed that the United States is ahead of China in most areas of technological competition, with the exception of 5G networks. The relative deterioration of the United States' position can be attributed to two factors: the lack of effective industrial policy over the past three decades in the United States and the relative improvement of China's position due to the size of its market and population. A brief overview of the EU position has also shown why the EU is struggling to create its own globally competitive technology

Table 1.6 Ranking of China, United States, and EU in AI competition

	China	United States	European Union
Talent	3	1	2
Research	3	1	2
Development	3	1	2
Adaption	1	3	2
Data	1	2	3
Hardware	2	1	3

Source: Castro et al. (2019).

firms and why this situation is unlikely to change in the future. The next chapter explores the question of what political and business motivations are driving changes in the EU.

Notes

1 The difference between geopolitics and geoeconomics is crucial; the latter has had the tendency to replace the original term in the absence of war in recent years. More on that in the third chapter of this book!
2 Gordon G. Chang' f. ex. predicted the imminent collapse of China's economic and political system in 2012 and 2016 (Chang, 2001), several times updating the exact year of the anticipated collapse.
3 China supported stopping the North Korean nuclear program and the India-Pakistan nuclear arms race. The country joined the Comprehensive Test Ban Treaty, the Chemical Weapons Convention, and claims to follow in principle the Missile Technology Control Regime. China has contributed to multilateral institutions (APEC, ASEAN, the World Bank, and the International Monetary Fund).
4 Chalmers Johnson coined the term "developmental state" in the 1980s. He tried to show the significant differences between capitalist countries by focusing on Japan. As he put it: "One of my purposes in introducing of the "capitalist developmental state" into a history of modern Japanese industrial policy was to go beyond the contrast between the American and Soviet economies" (Johnson, 1999, p. 32). He emphasized that the competent and farsighted bureaucracy greatly contributed to the Japanese economic miracle. Later, the concept "developmental state" became popular, and major contributions were made by Alice Amsden (1989), Robert Wade (1990), and others. Emphasis was put on other elements too: investment and various policy instruments (savings and credit schemes, foreign investment, export processing zones, government interventions to spread technology, etc.), history, and culture.
5 Chinese state-owned enterprises account for around 30 percent of the GDP (Zhang, 2019).
6 Karpov writes thus:

> Such state of affairs where central government retains the rights to set the prices of natural monopolies, capital and national currency, while local authorities retain the rights to design 'plan-market frontier', using tremendous variety of bargained 'tracks', I propose to call 'multiple-track price setting model'. Each 'track' is, in fact, a sum of conditions on which different units of the system participate in the Chinese domestic 'market'. This sum of

conditions for the concrete 'track' takes shape through non-transparent bargaining between this unit and corresponding level of party-state authorities or between mutually depending units under control and patronage of the corresponding party-state organs. Thus the 'tracks' are bargained between party-state organs of different levels, between enterprises (social units) and party-state organs and between enterprises (social units) themselves but under the party-state's auspices.

(Karpov, 2018: 120)

7 The paradigm of the developmental state is not the only attempt to frame and interpret the Chinese model. The term "Beijing Consensus"—a clear reference to the Washington Consensus—was invented by Joshua Ramo, which emphasized three crucial elements of Chinese success: the value of innovation, the rejection of the GDP-per-capita approach, and self-determination (Ramo, 2004: 11–12). Although the term became popular for a short time, it failed to reflect many other characteristics of Chinese economic development and contrast the Chinese experience with the example of Japan, South-Korea, Taiwan, and Singapore.

8 Levy and Fukuyama point out that there are five important elements that matter in the long run to establish a well-functioning political system: rule of law, rapid economic growth, democratic institutions, a competent and efficient state bureaucracy, and a vibrant, strong civil society. But they also argue that it is the sequence that matters—that democracy is not a necessary element of the whole catch-up process. Consider the examples of Japan and Germany, where the economy advanced much earlier than a stable democracy could be established (Levy & Fukuyama, 2001: 3). Not everyone agrees on this point: while Mansfield and Snyder are cautious about democratization without having an efficient (and impartial), relatively competent state mechanism, Carothers and Berman doubt that the right sequence (first state formation, then holding democratic elections) is necessary to reduce the risk of violence during the transition (Berman, 2007: 14–47; Carothers, 2007: 17–27). Based on the idea of sequencing, the crucial question remains: what is the possible scenario that China faces in its development? In the Chinese model, the formation of a more or less efficient state/bureaucracy is the first element, followed by robust economic growth, and then an emerging middle class. However, the rule of law and democratic procedures based on consultation are weak by Western standards. There is a clear process of modernization, as the Chinese model is more democratic than ever and is attempting to function more democratically, albeit selectively, at the local level and within the Communist Party itself. Goralczyk refers to Zheng Yongnian's work in Chinese, who maintains that China is in a phase of empowering its society. As in the case of Taiwan and South Korea, after this phase, the country will be ready to complete the democratization process (Goralczyk, 2017: 45). However, there is the example of Singapore, where, in addition to the former element, the rule of law was strongly implemented and certain democratic institutions were also in place, albeit not in the Westminster model of democracy. The question that cannot be answered at this stage is whether the Chinese elite can find ways to use some elements of the Singaporean experience to make the Chinese economic growth rate sustainable and self-sustaining.

9 Based on Word Bank's household consumption expenditure in current $ in year 2018.

10 The EU Chamber of Commerce and the US Chamber of Commerce also emphasized the import substitution nature of China's technology development plans in their 2017 reports (European Union Chamber of Commerce in China, 2017; US Chamber, 2017).

11 The Big Push theory, originally coined by Rosenstein-Rodan in 1943, argued that minor changes in fiscal and monetary policy do not suffice when expansionary and developmental policies are to be pursued. It proposed the large-scale investment of foreign aid in various sectors of the economy to accelerate development (Rosenstein-Rodan, 1943: 202–211).

12 Referring to Mao's ideas on scientific and technological progress, Julian Baird Gewirtz explains both how deeply technology is embedded in China's economic development strategy and that there is a strong link between technological strength and power:

> He [Mao] envisioned the socialist world's 'overwhelming superiority' in science and technology and came to see technological strength as central to economic, ideological, and geopolitical power—the view of catch up and surpass that CCP leaders continue to hold today.
>
> (Gewirtz, 2019)

13 The question remains why this strategy would be more efficient this time, why China could not repeat this success and rely more on alternative technologies. When the US blacklists Huawei and imposes restrictions on the import and export of Chinese technology, it will only accelerate China's domestic innovation processes. The measures introduced are backfiring by forcing Chinese firms to look for other alternatives, not only indigenous innovation, but also collaborations with other countries. This is the second point of the American strategy when it attempts to convince allies not to cooperate with China in technology development and not to open their markets to Chinese goods and services. At the same time, we should add that regardless of the American strategy, the next logical step in Chinese economic development is to become more independent and move up the value chain.

14 He says: "Top-down, CCP-led technological innovation brings its share of challenges. Many observers correctly cite the risks of misguided government-steered investment, which has led to waste and massive oversupply, or the challenges of supporting small entrepreneurs and researchers without heavy-handed interference" (Gewirtz, 2019).

15 The "Strategic Emerging Industries" was the 15-year plan adapted in 2006. It focused only on innovation, while "Made in China 2015" centered on all production. Unlike the previous strategy, "Made in China 2015" also targeted traditional industries and modern services, and there is more room for market forces in the document (Kennedy 2015).

16 Ferguson and Schularick, who coined the term "Chimerica" and tried to make a case for symbiosis, also wrote about the desirable end of Chimerica. They argued that the pegging of the Chinese yuan and the devaluation of the American dollar put pressure on Europe and Japan because "the burden of adjustment falls disproportionately on Europe and Japan" (Ferguson & Schularick, 2009: 4).

17 As the data is in purchasing power parity, price level differences are not relevant from this point of view.

18 According to the UN, around 28 percent of the global manufacturing output in 2018 was from China.

19 Their study analyzed six dimensions: (1) talent (the number of AI researchers, the number of top AI researchers based on the H-index, and attendance at academic conferences, etc.); (2) research (the number of AI papers, the field-weighted citation impact, the top 100 software and computer service companies for R&D spending, etc.) (3) development (number of AI start-ups, number of acquisitions, etc.); (4) adaptation (the number of workers using or piloting AI); (5) data (landlines, number of individuals using mobile phones, electronic health data, etc.); and (6) hardware (number of firms in the top 15 for semiconductor

sales, number of firms in the top 10 for semiconductor R&D spending, the number of firms designing AI chips, etc.). The unique feature of the analysis is that the authors of the paper adjusted the numbers, and thus their ranking, based on the number of workers or individuals.

References

ABI Research (2019, October 3). China's AI Ambition Gets a Reality Check as the USA Reclaims Top Spot in Global AI Investment. *ABI Research*, Retrieved from: https://www.abiresearch.com/press/chinas-ai-ambition-gets-a-reality-check-as-the-usa-reclaims-top-spot-in-global-ai-investment/

Abrami, R. M., Kirby, W. C. & McFarlan, F. W. (2014, March). Why China Can't Innovate. *Harvard Business Review, 3*(92), pp. 107–111.

Acemoglu, D. & Robinson, J. A. (2012). *Why Nations Fail?* New York: Crown Publishers.

Amsden, A. H. (1989). *Asia's Next Giant: South Korea and Late Industrialization.* New York and Oxford: Oxford University Press.

Bain and Company (2019). Global Private Equity Report 2019, Retrieved from: https://www.bain.com/contentassets/875a49e26e9c4775942ec5b86084df0a/bain_report_private_equity_report_2019.pdf

Bello, W. (2017, February). Keynesianism in the Great Recession. Right Diagnosis, Wrong Cure. *Transnational Institute.* Finance Working Papers, Global Finance Series, Retrieved from: https://www.tni.org/en/publication/keynesianism-in-the-great-recession

Benner, K. (2020, February 6). China's Dominance of 5G Networks Puts U.S. Economic Future at Stake, Barr Warns. *New York Times*, Retrieved from: https://www.nytimes.com/2020/02/06/us/politics/barr-5g.html

Berman, S. (2007, January). The Vain Hope for Correct Timing. *Journal of Democracy, 18*(3), pp. 14–17.

Carothers, T. (2007, January). The Sequencing Fallacy. *Journal of Democracy, 18*(1), pp. 12–27.

Castro, D., McLaughlin, M. & Chivot, E. (2019, August 19). Who Is Winning the AI Race: China, the EU or the United States? *Center for Data Innovation*, Retrieved from: https://datainnovation.org/2019/08/who-is-winning-the-ai-race-china-the-eu-or-the-united-states/

Chang, G. G. (2001). *The Coming Collapse of China.* New York: Random House.

Council on Foreign Relations (2019). Innovation and National Security Keeping Our Edge. Independent Task Force Report No. 77, Retrieved from: https://www.cfr.org/report/keeping-our-edge/pdf/TFR_Innovation_Strategy.pdf

Davies, J. (2020, August 4). Nokia, Ericsson and Huawei Dominance Beginning to Fade—Analyst. *Informa PLC*, Retrieved from: https://telecoms.com/505872/nokia-ericsson-and-huawei-dominance-beginning-to-fade-analyst/

Divine, J. (2020, September 17). The 10 Most Valuable Tech Companies in the World. *U.S. News*, Retrieved from: https://money.usnews.com/investing/stock-market-news/slideshows/most-valuable-tech-companies-in-the-world

Economy, E. C. (2018). *The Third Revolution. Xi Jinping and the New Chinese State.* New York: Oxford University Press.

Economy, E. C. (2019, March 6). The Problem with Xi's China Model. Why Its Successes Are Becoming Liabilities. *Foreign Affairs*, Retrieved from: https://www.foreignaffairs.com/articles/china/2019-03-06/problem-xis-china-model

European Commission (2020). EU R&D Scoreboard. The 2020 EU Industrial R&D Investment Scoreboard, Retrieved from: https://iri.jrc.ec.europa.eu/scoreboard/2020-eu-industrial-rd-investment-scoreboard

European Union Chamber of Commerce in China (2017). *China Manufacturing 2025*. Putting Industrial Policy Ahead of Market Forces.

Federal Ministry for Economic Affairs and Energy (2019). Industrial Strategy 2030. Strategic Guidelines for a German and European Industrial Policy, Retrieved from: https://www.bmwi.de/Redaktion/EN/Publikationen/Industry/national-industry-strategy-2030.html

Ferguson, N. (2011). *The West and the Rest*. London: The Penguin Press.

Ferguson, N. & Schularick, M. (2009). The End of Chimerica. *Harvard Business School*. Working Paper 10–037, Retrieved from: https://www.hbs.edu/faculty/Publication%20Files/10-037_0fdf7d5e-ce9e-45d8-9429-84f8047db65b.pdf

Fitzwater, M. (1991). Statement by Press Secretary Fitzwater on Restrictions on U.S. Satellite Component Export. In Public Papers of the Presidents of the United States of America, Retrieved from: https://www.govinfo.gov/content/pkg/PPP-1991-book1/pdf/PPP-1991-book1.pdf

Forbes (2020, May 13). Global 2000. The World's Largest Public Companies. *Forbes*, Retrieved from: https://www.forbes.com/global2000/#7b3aa032335d

Gewirtz, J. B. (2019, August 27). China's Long March to Technological Supremacy: The Roots of Xi Jinping's Ambition to "Catch Up and Surpass." *Foreign Affairs*, Retrieved from: https://www.foreignaffairs.com/articles/china/2019-08-27/chinas-long-march-technological-supremacy

Goralczyk, B. (2017). Eastern-Asian Development Model: The Growth-Fostering State. In Lin, Y. J. & Nowak, A. Z. (Eds.). *New Structural Economics for Less Advanced Countries*. Warsaw: University of Warsaw, Faculty of Management Press, pp. 41–55.

GreyB (2020). Who Owns Core 5G Patents? A Detailed Analysis of 5G SEPs. *GreyB and Amplified*, Retrieved from: https://www.greyb.com/5g-patents/

Johnson, C. (1999). The Developmental State: Odyssey of a Concept. In Woo-Cumings, M. (Ed.). *The Developmental State*. Ithaca: Cornell University Press, pp. 32–60.

Karpov, M. (2018). China's Institutional "Miracle"—Party-State in the Transition to Market Economy: Potential and Limits of Systemic Sustainability. In Moldicz, C. (Ed.). *Dilemmas and Challenges of the Chinese Economy in the 21st Century: Economic Policy Effects of the Belt and Road Initiative*. Budapest: Oriental Business and Innovation Center (OBIC)/Budapest Business School, pp. 107–135.

Kasahara, S. (2013, November). *The Asian Developmental State and the Flying Geese Paradigm*. United Nations' Conference on Trade and Development, Discussion Papers, Retrieved from: https://unctad.org/system/files/official-document/osgdp20133_en.pdf

Kennedy, S. (2015, June 1). *Made in China 2015*. Center for Strategic and International Studies, Retrieved from: https://www.csis.org/analysis/made-china-2025

Kristof, D. N. (1993, November/December). The Rise of China. *Foreign Affairs*, Retrieved from: https://www.foreignaffairs.com/articles/asia/1993-12-01/rise-china

Laskai, L. (2019, June 19). Why Blacklisting Huawei Could Backfire. The History of Chinese Indigenous Innovation. *Foreign Affairs*, Retrieved from: https://www.foreignaffairs.com/articles/china/2019-06-19/why-blacklisting-huawei-could-backfire

Levy, B. & Fukuyama, F. (2010). Development Strategies. Integration Governance and Growth. *The World Bank*. Policy Research Working Paper No. 5196, Retrieved from: https://openknowledge.worldbank.org/bitstream/handle/10986/19915/WPS5196.pdf?sequence=1&isAllowed=y

Lipset, S. M. (1959, March). Some Social Requisites of Democracy: Economic Development and Political Legitimacy. *The American Political Science Review, 53*(1), pp. 69–105. Retrieved from: https://scholar.harvard.edu/files/levitsky/files/lipset_1959.pdf

Mackinder, J. H. (1919). *Democratic Ideals and Reality: A Study in the Politics of Reconstruction*. New York: Henry Holt and Company, p. 150.

Maincent, E. & Navarro, L. (2006, April). A Policy for Industrial Champions: From Picking Winners to Fostering Excellence and the Growth of Firms. *Enterprise and Industry Directorate-General European Commission*. Industrial Policy and Economic Reforms Papers No. 2, Retrieved from: https://op.europa.eu/en/publication-detail/-/publication/768c3d1a-ae43-479c-b2cb-460248d9f910

Mead, W. R. (2014, May/June). The Return of Geopolitics. The Revenge of the Revisionist Powers. *Foreign Affairs*, Retrieved from: https://www.foreignaffairs.com/articles/china/2014-04-17/return-geopolitics

National Counterintelligence and Security Center (2020). National Counterintelligence Strategy of the United States of America 2020–2022, Retrieved from: https://www.dni.gov/index.php/ncsc-features/2741-the-national-counterintelligence-strategy-of-the-united-states-of-america-2020-2020

National Science Board (2018). Science and Engineering Indicators 2018, Retrieved from: https://www.nsf.gov/statistics/2018/nsb20181/assets/nsb20181.pdf

National Science Board (2020, January 15). Science and Engineering Indicators 2020. *Research and Development: U.S. Trends and International Comparisons*. NSB-2020–3, Retrieved from: https://ncses.nsf.gov/pubs/nsb20203

Naughton, B. (2007). *The Chinese Economy. Transitions and Growth*. Cambridge/London: The MIT Press.

Neil, T. (2019, September 3). Matters of Record: Relitigating Engagement with China. *Marco Polo*, Retrieved from: https://macropolo.org/analysis/china-us-engagement-policy/

Pantzov, A. V. & Levine, S. I. (2012). *Mao: The Real Story*. New York: Simon & Schuster Paperbacks.

Pompeo, M. R. (2020, June 25). *A New Transatlantic Dialogue. Speech.* Washington: German Marshall Fund's Brussels Forum, Retrieved from: https://useu.usmission.gov/secretary-pompeos-remarks-at-brussels-forum/

Pongratz, S. (2020, March 2). Huawei and ZTE Increased Their Revenue Shares While Nokia and Cisco's Revenue Shares Declined for the Full Year 2019 Telecom Equipment Market. *Dell'Oro Group*, Retrieved from: https://www.delloro.com/the-telecom-equipment-market-2019/

Ramo, J. C. (2004). The Beijing Consensus. *The Foreign Policy Centre*, Retrieved from: https://fpc.org.uk/wp-content/uploads/2006/09/244.pdf

Rosenstein-Rodan, P. N. (1943). Problems of Industrialization of Eastern and South-Eastern Europe. *Economic Journal, 53*(210/211), pp. 202–211. Retrieved from: https://www.econ.nyu.edu/user/debraj/Courses/Readings/RosensteinRodan.pdf

Sawers, P. (2020, March 16). Tech Nation: U.S. Companies Raised 56% of Global AI Investment since 2015, Followed by China and the U.K. *VentureBeat*,

Retrieved from: https://venturebeat.com/2020/03/16/tech-nation-u-s-companies-raised-56-of-global-ai-investment-since-2015-followed-by-china-and-u-k/

Select Committee, U.S. House of Representatives (1999). Concerns with the People's Republic of China, U.S. National Security and Military/Commercial, Rep. Christopher Cox, Chairman. Washington: U.S. Government Printing Office.

Setser, B. W. (2019, August 8). Is China Manipulating Its Currency? *Council on Foreign Relations*, Retrieved from: https://www.cfr.org/in-brief/china-manipulating-its-currency

Shih, W. C. (2020, April 15). Bringing Manufacturing Back to the U.S. Is Easier Said Than Done. *Harvard Business Review*, Retrieved from: https://hbr.org/2020/04/bringing-manufacturing-back-to-the-u-s-is-easier-said-than-done

Sun, H. (2019). U.S.-China Tech War. Impacts and Prospects. *China Quarterly of International Strategic Studies, 5*(2), pp. 197–212.

U.S. Chamber of Commerce (2017, March). Made in China: 2025. Global Ambitions Built on Local Protections. *U.S. Chamber of Commerce*, Retrieved from: https://www.uschamber.com/sites/default/files/final_made_in_china_2025_report_full.pdf

Wade, R. (1990). *Governing the Market: Economic Theory and the Role of Government in East Asian Industrialization*. Princeton: Princeton University Press.

Weiss, K. (2019, September 6). As Britain Prepares to Leave, the EU Slides Further Toward Protectionism. *CapX*, Retrieved from: https://capx.co/as-britain-prepares-to-leave-the-eu-slides-further-towards-protectionism/

WIPO (2020, April 7). China Becomes Top Filer of International Patents in 2019 Amid Robust Growth for WIPO's IP Services, Treaties and Finances. *WIPO*, Geneva. R/2020/848, Retrieved from: https://www.wipo.int/pressroom/en/articles/2020/article_0005.html

World Bank (2020). World Bank WITS Database, Retrieved from: https://wits.worldbank.org/

Ye, J. (2002). Will China Be a Threat to Its Neighbors and the World in the Twenty First Century?. *Ritsumeikan University, Annual Review of International Studies, 1*, p. 57.

Zhang, Z. Z. (2019, May 19). China's SOE Reforms: What the Latest Round of Reforms Mean for the Market. *China Briefing*, Retrieved from: https://www.china-briefing.com/news/chinas-soe-reform-process/

2 The war of arguments

The European battlefield

2.1 EU-China debates

After the establishment of diplomatic relations between the People's Republic of China (PRC) and the European Economic Community in 1975, diplomatic and economic relations initially developed rather slowly. It took nine years for the first ministerial-level meetings to take place and 13 years for the Beijing Delegation of the European Commission (EC) to open in 1988. After the setback of the Tiananmen Square Protests, a new round of bilateral dialogue was launched in 1992. Since 1998, the EU-China Summits have laid the foundation for the EU-China relationship. The last EU-China Summit, the 22nd, took place in 2020, held by video conference due to the coronavirus (European Council, 2020).

Before we delve deeper into EU-China relations, let us not forget that EU members often have their own China strategies. It may sound trivial at first, but one aspect that is often overlooked is that the EU's China strategy often collides with EU member states' perceptions of China, not to mention other European countries' perceptions of China. What is the main reason for this collision? The different economic development interests of Western and Central European countries complicate a unified China policy and strategy within the EU. Not only the different levels of economic development, but also the different corporate structures in Western and Central Europe cause decision-makers to come up with very different responses.[1] This can be illustrated by a simple example: regional supply chains in Europe are organized by German, French, and Italian firms, not by Central European firms. If Chinese firms enter the Single Market via foreign direct investments (FDIs), Western European countries have more to lose than Central Europe. At the same time, the Single Market and EU-level policies provide a common framework for the policies implemented by each country in the EU. This sounds like a recipe for disaster: common rule and different economic interests.[2]

If there is no agreement among European countries on a consistent China strategy, it is not surprising at all that there is clear disagreement between the United States and its European allies on a coherent China perception

DOI: 10.4324/9781003128625-2

and strategy. The so-called trade war has only deepened the dividing lines between the United States and Europe in recent years, making it more difficult to implement a common China policy in the West (Small, 2019). Differing American and European interests are clear, as the EU has no explicit geopolitical interests in East Asia while the United States seems to be pursuing a new "containment" policy toward China. However, EC and European Parliament views on China have hardened recently. Small argued in 2019 that while political and security developments have played an important role in the change, economic aspects have arguably been more significant. He stated:

> Europe has lost hope that China will reform its economy or allow greater access to its markets, and at the same time, China's state-backed and state-subsidized actors have advanced in sectors that Europe considers critical to its economic future.
>
> (Small, 2019)

The term "Europe" used by Small is not entirely accurate: it most likely refers to the changing attitudes of Germany and France toward China, or those of the European institutions reflected in the last EC paper, which first used the term "systemic rival" for China:

> China is, simultaneously, in different policy areas, a cooperation partner with whom the EU has closely aligned objectives, a negotiating partner with whom the EU needs to find a balance of interests, an economic competitor in the pursuit of technological leadership, and a systemic rival promoting alternative models of governance. This requires a flexible and pragmatic whole-of-EU approach enabling a principled defence of interests and values.
>
> (European Commission, 2019a: 1)

With the introduction of the notion "systemic rival," the European approach is very far from the more optimistic attitude toward China that characterized bilateral relations when the EU declared China one of the six strategic partners in its very first security strategy in 2003 (Council of the European Union, 2003). In the same year, the two countries announced a strategic partnership, and the first full revision of this relationship took place in 2016, when the "EU-China 2020 Strategic Agenda for Cooperation" was adopted by China and the EU. The approach was updated in the document cited above, which is the latest comprehensive China strategy at EU level.

The strategy was issued by the EC in 2019 (European Commission, 2019a), but since the inception of the new EC in late 2019, the tone of the approach to China seems to be more hostile. After the 10th annual Strategic Dialogue between the EU and China, which paved the way for the 22nd summit, Josep Borrell, the EU High Representative for Foreign Affairs, said in a press conference referring to the previous summit that there are still problems that need to be solved; market access, a level playing field, and reciprocity are the main commitments from the Chinese side that EU would like to see Beijing implement. In the next two parts of this chapter, we analyze the different levels of issues by dividing them into "economic development debates" and "political debates."

2.1.1 Economic development debates

These debates originate from different approaches to how to develop the economy. China has a developmental state approach that is quite different from the model accepted in the West, but very similar to what was seen in the early days of the Asian developmental states. Even then, the different economic development paradigm in these East Asian countries seemed unacceptable to many observers in the West. Bergsten summarized the debate between the United States and Japan in 1982 thus:

> To be sure, there has been fairly steady tension between the United States and Japan over economic issues ever since Japan emerged as a major industrial power. Japan's amazing success ... has won its grudging admiration but also growing hostility as a disruptive force in American economic life and brought repeated charges of 'unfair' competition. Its apparent reluctance, or even inability, to expand substantially its imports of manufactured products has produced steady charges that Japan is itself highly protectionist, a 'free rider' on the open trading system from which it benefits so greatly but within which it seems unwilling to provide others with truly reciprocal opportunities.
>
> (Bergsten, 1982: 1059)

Pursuit of a protectionist trade policy and use of unfair trade practices, reluctance to open the domestic market for foreign firms; these were the arguments used against Japan here, and we may already be familiar with them from the EU-China debate. The same issues were framed slightly differently in the EU press release following the 22nd EU-China Summit (European Commission, 2020, June 22), but the same critical points were raised that we can recognize in previous debates between the United States and Japan and between the United States and Germany. The press release highlights several unresolved issues in EU-China relations. In the remarks, problems in the Chinese market and the Single Market of the EU are addressed. When problems in access to the Chinese market are concerned, the EC pointed to problems such as asymmetries in market access; level playing field for European firms in the Chinese market (including the problem of industrial subsidies to be regulated by the World Trade Organization (WTO) mechanisms, overcapacity problems in traditional sectors such as steel, but also in high-tech areas), and forced technology transfers.

These issues were partially resolved in the EU-China Comprehensive Investment Agreement in December 2020. Considering that the negotiations started back in 2014 and that no agreement was reached for seven years, one can understand that the difficulties that the negotiators had to solve were not of a technical nature, but fundamental differences between the two economic regimes had to be addressed during the negotiations. Aware of the complexity of the issues, the partners agreed in 2016 that the investment agreement should go beyond the framework of traditional investment agreements and include market access commitments. In our understanding, it was extremely difficult to achieve the goal of a comprehensive agreement, as the public sector reforms demanded by the EU touched the core of China's development model and risked changing it completely.[3]

However, before evaluating the achievements of the EU-China Comprehensive Investment Agreement, one should briefly test the validity of the EU arguments in the debate leading to concluding the agreement as access to market and level playing field are concerned.

Looking at Table 2.1, which shows FDI flows between 2014 and 2019, we find that recent Chinese FDI in the EU exceeded EU FDI in China, with the investment cycle peaking in 2016 at $37.3 billion and 2017 at $29.2 billion. EU countries' FDI in China fluctuated between $7 billion and $13 billion between 2014 and 2019. Nonetheless, these sums are significantly lower, but European FDI in China has a much longer history, beginning in the late 1990s, while Chinese FDI only started to flow into the EU in significant amounts after the Global Financial Crisis (2008–2009). Due to this time gap between EU FDI and Chinese FDI, the two FDI stocks are close to each other today. Based on these figures, it can be concluded that although European FDI is more constrained in the Chinese economy, the size of the Chinese economy can compensate for European companies and a balanced situation could be achieved at least at macroeconomic level.

At the same time, in the trade with China, market access of European companies is limited, and we are not even close to the reciprocity we see in investment relations at the macroeconomic level. The latest Global Enabling Trade Report confirms this European criticism (World Economic Forum, 2016); China ranks 61st on this list, while European trading partners are in the first part of the ranking.[4] The index is composed of four subindices: market access, border administration, transport and communication

Table 2.1 FDI flows between China and the EU (€ billion)

	2014	2015	2016	2017	2018	2019
EU FDI in China	13.0	10.0	8.0	6.8	7.0	13.0
Chinese FDI in EU	14.7	20.7	37.3	29.2	17.4	11.7

Sources: Kratz et al. (2020), Rhodium Group. Cross Border Monitors. People's Republic of China and the European Union.

Table 2.2 Several EU countries ranked based on the size of the trade deficit with China (2018; € billion)

	The balance in China trade	The overall trade balance
France	−34	−59
Netherlands	−33	55
Poland	−28	−6
Spain	−24	−48
Italy	−20	46
Germany	−16	270

Source: Own compilation based on World Bank WITS database.

infrastructure, and business environment. The market index has two further subcategories: foreign market and domestic market access. The latter is important for our analysis; here the country ranks 101st in the list of 136 countries. As a result of the uneven playing field, China had a significant trade surplus (2019: €164 billion) in goods with the EU, which cannot be offset by a trade deficit in services (2018: €17 billion). At this point, we must add that the Chinese trade surplus can be explained not only by an uneven playing field, but also by significant competitive advantages in manufacturing. If we look at Table 2.2, we can see the trade balance of the larger European economies with China and even their overall trade balance, which helps us understand why the trade deficit is so painful (France, Poland, Spain) and why it is less important (Germany, Netherlands, Italy).

Forced transfer of technology is the second key complaint of European and American firms in China, and more and more firms are reporting such cases. We can better understand the present and better predict future developments by looking to the past. In this debate, we often forget that ignoring intellectual property rights can be a sound policy choice, as it was when the United States was the largest infringer of intellectual property rights in the 19th century. Pang at al. draw our attention to the gap between existing, well-developed law and its enforcement in China. At the same time, they also conclude that China will change its behavior if its own rights can be violated by other countries:

> Leveraging this period of US history, we predict that to the same extent that the United States voluntarily agreed to strengthen IPR protection when the US economy became sufficiently innovation-driven, China will similarly enhance its IPR protection. We further predict that when Chinese IPR are significantly violated abroad, China will become more serious about IPR protection.
>
> (Peng et al., 2017)

Recent data on forced transfer of technology is not conclusive about whether China has changed its attitude toward intellectual property rights in any direction. According to the 2020 annual survey by the EU Chamber of Commerce, 16 percent of foreign firms were forced to hand over their technology, a significantly higher figure than the 10 percent reported in 2017 (European Union Chamber of Commerce in China, 2020: 31), but a drop from the 20 percent reported in 2019.

We argued in the first chapter that forced technology transfer can be a reasonable choice when it comes to developing countries. Spross argues that technology transfer can be interpreted as the price of entry into China's vast market.[5] However, he adds that it is not the only way, but it is certainly a way to level the playing field between advanced and developing countries. In the case of China, he questions the need for the upper middle-income country to still rely on this practice (Spross, 2019).

Problems in the Single Market originate mainly from the behavior of state-owned enterprises and the transparency of Chinese state subsidies. The existing tensions between the EU and China have been created partly by the slow development of EU law and partly by a new wave of globalization that, unlike earlier globalization waves, includes state-owned enterprises as well. Chinese state-owned enterprises entered the European market after 2010–2011 and became increasingly involved in acquisitions. It was during these years that the EU first became alarmed about the dangers involved.

Under existing rules, state aid and state-owned enterprises of the member states are monitored by the EC, as the absence of state aid rules would distort the Single Market. This is precisely why EU competition policy was one of the first common policies in the EU. At that time, it was less typical for foreign state-owned companies or companies with significant state support (cheap loans, capital, or guarantees) to enter the Single Market, so the body of law did not have specific rules for such cases. However, after 2008–2009, an increase in transactions with foreign state-owned or state-supported companies was registered, which led to a new situation in recent years. The EC responded to this problem by adopting a White Paper on June 17, 2020 (European Commission, 2020a). The White Paper launched a public consultation on this issue until September 23, 2020. The proposal focuses on three key areas: (1) on the Single Market in general, (2) on the acquisition of EU members facilitated by foreign subsidies, and (3) on foreign subsidies in EU public procurement procedures. The aim is to prepare appropriate legislative proposals in this area; according to the plans of the EC, a legislative proposal would come forward in 2021.[6]

A look at the direct investment data shows that the European response is probably too late and insufficient, as the share of Chinese state-owned enterprises in total Chinese FDI to the EU, which was over 70 percent between 2010 and 2015, has fallen to 7 percent in 2019 (Kratz et al., 2020). This decline in the share of state-owned enterprises is global, which can be explained by an overall, more cautious attitude toward Chinese investment in the world. Another explanation has been offered by Derek Scissors, which draws attention to declining Chinese foreign exchange reserves leading to significant changes in China's investment behavior (Scissors, 2019).

In the long run, our understanding is that the share will not grow because the Chinese private sector is dynamic enough to compensate for the change in behavior of state-directed enterprises; research and development (R&D) in particular are dominated by the private sector too. Around 60 percent of GDP is produced by private firms, and 70 percent of innovation comes from this part of the economy (Guluzade, 2019).

It must be added that private firms can also benefit from lavish government subsidies that distort the market. Regulating these activities would require global cooperation at best, which seems highly unlikely at present, but the EU-China Comprehensive Investment Agreement promises to ease this pain as well as transparency of Chinese state subsidies are concerned. This is a less focused aspect of the debate, but global solutions are not currently in sight. For this reason, the general impression is that China is dominating with its direct investment and buying up the world. Let's take a look and ask the question whether China is investing "too much" in the world and the EU, or in other words, whether Chinese direct investment in the world is in proportion to foreign investment in China:

1   Using the World Bank database, we can see that Chinese FDI abroad and foreign FDI in China seem to be roughly proportional; Chinese FDI stock abroad was 15 percent of GDP, while FDI at home was 20 percent of the country's GDP in 2019.

2   The question could also be raised whether Chinese FDI in the EU is disproportionate to EU investment in China, somehow violating the principle of reciprocity—often cited by European policymakers. Based on the Eurostat database, China (including investment from Macau and Hong Kong) was responsible for a 4.0 percent share of the EU-27 economy's outward FDI positions at the end of 2019, and 4.0 percent of the EU-27s outward FDI stocks were held in China. From this perspective, it is rather a balanced relation.

3   We can ask how the Chinese positions in the EU look like if we compare them with their competitors. The so-called traditional foreign investors (the United States, Canada, Switzerland, Norway, Australia, and Japan) dominate this segment, controlling more than 80 percent of the foreign-owned assets in the EU-28 (European Commission, 2019b: 11). So we can conclude that China's role in FDI in the EU is much smaller than might be assumed from the size of its economy or its role in trade.

4   China (including investment from Macau and Hong Kong) has rapidly increased its share of businesses in the EU-28. In 2017, China's share of non-EU controlled companies was 9.5 percent and held 3 percent of all non-EU controlled assets.[7] To grasp the significance of this figure, we should add that about 2.8 percent of all firms located in Europe were

Table 2.3 Foreign firms share in EU-28 (number of firms and share of assets)

	Foreign firms share (%)	China's share within the subset of foreign firms (Macao, Hong Kong) (%)	United States and Canada's share within the subset of foreign firms (%)	China's share within the subset of EU firms (Macao, Hong Kong)	The United States and Canada's within the subset of EU firms
Number of firms	2.8	9.5	29	2.94	9.65
Share of assets	35	3.0	61.8	0.86	17.65

Source: European Commission (2019b: 11).

owned by foreigners, who held 32.8 percent of the assets. In other words, we are talking about the 9.5 percent of the 2.8 percent and 3 percent of the 32.8 percent. In stark contrast to the general interpretation, these shares do not seem to be relevant (see Table 2.3).

2.1.2 Political debates

The above-mentioned EC press release lists some unresolved issues between the EU and China that require special attention and concern aspects of both economic development and social development (European Commission, 2020, June 22): the link between digital technologies and respect for fundamental rights and data protection; cybersecurity and disinformation and issues related to amendments in the Hong Kong Basic Law; and the state of the human rights situation, including the treatment of minorities in Xinjiang and Tibet.

The fundamental question is whether linking values to economy-related issues can produce long-term results and bring China's political system closer to Western political institutions. If the answer is yes, it makes sense to follow the United States and enforce a "democracy export" foreign policy like the United States, but as we understand it several factors argue that the pursuit of a pragmatic foreign policy would make more sense in general, and in the case of China:

* the lack of geopolitical interest and hard power in the Asian region;
* the inefficiency of democracy export;
* the respect for local values; and
* the historical experiences of China.

Unlike the United States, the EU as a whole has *no geopolitical interest* in the region—only economic interests, and these are best served by focusing on defending European business interests. France is the only country in the EU which has a potential clash of interests with China due its traditional sphere of influence where China's clout has risen considerable in the last decade. Even if we assume that the EU has political interests in the Asia-Pacific region, it *lacks the hard power* to enforce those interests. The soft power of Europe on China is significant: the culture, languages, traditions, and especially the lifestyle in these regions are attractive to the Chinese.

The last few decades, especially the recent experience of American foreign policy, have shown how *useless democracy export* can be when it is imposed on a society while large segments of society simply do not support the spread of Western-type institutions for various reasons (be they cultural, religious, or social). In the case of China, it is worth remembering that the country experienced an average annual GDP growth of 9.4 percent between 1980 and 2019. No wonder existing political and economic institutions can rely on the strong support of China's burgeoning middle class. Not only China, but much smaller societies have been very successful in "protecting" themselves from the establishment of Western-type institutions over the past decades. Even when they could be established by relying on small local elites, they were short-lived and soon distorted or reshaped to meet the needs of local society (Afghanistan, Yemen, and Iraq are good examples of these failed attempts). Mead sums up the end of this dream thus:

> The Wilsonian project requires a high degree of convergence to succeed; the member states of a Wilsonian order must be democratic, and they must be willing and able to conduct their international relations within liberal multilateral institutions. At least for the medium term, the belief in convergence can no longer be sustained. Today, China, India, Russia, and Turkey all seem less likely to converge on liberal democracy than they did in 1990.
>
> (Mead, 2021: 131)

As for China, *the perception of Chinese history* is an often-neglected factor among Europeans and other Westerners. Chinese history in the 19th and early 20th centuries is usually described as the century of humiliation, when Chinese were dictated to and advised by Western powers on how to organize their society and shape political and economic institutions. Even if this pressure was applied with the best of intentions, it was not until the late 1970s that the Chinese economy took off. The results came from its own ideas and solutions, but in the early stages it relied heavily on foreign capital and technology. China's economic success is reflected in the support of China's burgeoning middle class, which is the key to any political change or adjustment in China.

On the basis of "realpolitik" approach, the policy of noninterference seems more advisable; however, EU institutions feel the urge to engage China in ways that could lead to profound changes in the political regime of China. The best example of this approach is the European Parliament and

its members, who on the one hand can influence the agenda and tone of negotiations with China, and on the other hand whose reactions are motivated
by domestic or EU politics rather than the hard rationality of foreign policy.

In 1980, five years after the establishment of diplomatic relations with Beijing, the European Parliament set up a so-called Delegation for Relations
with the PRC (hereafter D-CN). Based on several communiqués issued by
the D-CN in recent years, the Delegation focuses mainly on the political
aspects of relations with the PRC and much less on the economic aspects.
These documents center on the new national security law in Hong Kong,
problems in West China, and tensions at the Sino-Indian border. A communiqué usually leads to EP resolutions, which was the case here. The European Parliament's March 12, 2019, resolution on security threats related to
China's increasing technological presence in the EU and possible EU-level
action to reduce them called on member states:

> …to inform the Commission of any national measure they intend to
> adopt in order to coordinate the Union's response so as to ensure the
> highest standards of cyber security throughout the Union, …
>
> (European Parliament, 2019)

And it called on the Commission:

> … to assess the robustness of the Union's legal framework in order to
> address concerns about the presence of vulnerable equipment in strate-
> gic sectors and backbone infrastructure; …
>
> (European Parliament, 2019)

The call resulted first in the EC's Recommendation on Cybersecurity of 5G
networks (European Commission, 2019b), then a so-called EU toolbox for
member states to help mitigate cybersecurity risks (European Commission,
2020b). (See more on the development of the European regulatory framework in Table 2.4.)

What we need to understand at this point is that the supranational institutions of the EU (the EC and the European Parliament) do not have exclusive
rights in regulating 5G and other technology-related policies. Rather, these
powers are at the level of individual members when it comes to national security. As we have argued, a lack of industrial policy has prevented the EU
from creating European champions and similarly 5G is regulated by member states. The press release for the launch of 5G puts it this way:

> While market players are largely responsible for the secure rollout of
> 5G, and Member States are responsible for national security, 5G net-
> work security is an issue of strategic importance for the entire Single
> Market and the EU's technological sovereignty.
>
> (European Commission, January 29, 2020)

It is worth paying attention to the phrase "technological sovereignty," which is a new term used by the EC. The term, along with the notion of China as a "systemic rival," is a new way of approaching and managing relations with China. The term "systemic rival" was first used in an EC communication in March 2019. Josep Borrell, the EU's High Representative for Foreign Affairs, explained the two elements of the term this way—first the word systemic, then rivalry:

> It is clear that we do not have the same political system. It is clear that China defends its political system as we do with ours. It is clear that China has a global ambition. But, at the same time, I do not think that China is playing a role that can threaten world peace. They committed once and again to the fact that they want to be present in the world and play a global role, but they do not have military ambitions and they do not want to use force and participate in military conflicts. What do we mean by 'rivalry'? Well, let's go over this word. Sometimes, there are differences in interests and values. That is a fact of life. It is also a fact of life that we have to cooperate because you cannot imagine how we can solve the climate challenge without strong cooperation with China. You cannot build a multilateral world without China participating in it effectively, not in a 'Chinese way,' but in a way that can be accepted by everybody.
>
> (Borell, 2020)

Borell's approach reflects a "realpolitik" that is as far away from the idealistic approach of the early 2000s and the bellicose tone used in recent American foreign policy communications. We argue at this point that if the "realpolitik" approach can be implemented in EU foreign policy, it can truly serve as a solid foundation for an effective foreign policy toward China. In the next subsection, we will see why and how one can argue for this more cooperative approach toward China.

2.2 A summary of the EU's position on China

European problems regarding entry and access to the Chinese market are real, they are relevant in trade and technology transfer, but they are less emphasized in FDI. Trade protectionism, heavy-handed state guidance (state subsidizes), and the violation of intellectual rights are inherent features of the economic development policies of emerging economies, whether it is the practice of the Asian developmental state (Taiwan, Japan, South Korea in their early stages of development) or the trade protectionism of the emerging United States or Germany at the end of the 19th century. Thus, we can conclude that this (Chinese) economic policy will not simply disappear, as it is an inherent element of the Chinese model; however, the negative features of the model can be slowly changed as development progresses.

When it comes to problems caused by Chinese firms in the Single Market, they seem to be exaggerated by policymakers because neither the size of Chinese FDI nor the ownership structure suffices to distort the Single Market. However, what can be seen as a significant problem is the lack of an EU industrial policy that could foster European champions. The reasons why the inflow of Chinese FDI is discussed in such a heated and exaggerated manner are due to the following factors:

(1) Most of the investment is concentrated in four countries: the United Kingdom,[8] Germany, Italy, and France, so these countries have more first-hand experience in dealing with Chinese firms while setting the tone of the European-China discourse.

(2) Germany is the only case where investments systematically target top technology-oriented companies in the automotive industry, which could actually hurt the core interests of German industry. That is why the change in European attitude toward Chinese FDI in recent years has emanated from Germany initially, intensified by France's suffering from trade deficit with China and fueled by the United States for foreign policy reasons.

Despite the hostile environment, the bulk of these highly debated questions can be managed under the framework of the EU-China Comprehensive Agreement on Investment (CAI). However, it took a lot of patience on the European side and more flexibility on the Chinese side to reach an agreement. The European institutions' understanding that compromises cannot be reached with China on every issue in the short or medium term was initially missing from the debate. But the fact that they were able to reach a political compromise on the EU-China CAI in December 2020 shows that business interests and the "realpolitik" approach dominated the European discourse on China ultimately.

As China was negotiating not only trade and investment but also its own economic and political structures, the necessary solutions to the problem were often difficult and time-consuming to find. Negotiations between China and the EU were complicated by the following constant factors:

1   China negotiated its political and economic system, which is why it was so difficult for China to adopt a give-and-take policy.
2   The EU has a Common Foreign and Security Policy (CFSP) but no unified foreign policy. Since the interests of member states are different, the coordination of these foreign policies is not as sufficient.
3   A realistic foreign policy approach to relations with China could be worked out; however, some political actors (members of the European Parliament, opposition parties) who can influence this relationship had other than foreign policy goals in mind.

According to the political deal reached in December 2020, the agreement regulates FDI, forced technology transfer, obligations for state-owned enterprises, transparency rules for state subsidies, and commitments for sustainable development.

As far as forced technology transfer is concerned, the CAI contains clear rules against this practice: it abolishes the requirements for forced technology transfer to a joint partner and prohibits interference with contractual freedom in technology licensing. Since WTO rules do not cover subsidies in services, the CAI fills this vacuum and provides rules for them.

The agreement also addresses labor and environmental protection, as China commits not to lower existing standards in these areas, to work on ratifying the International Labor Organization conventions, especially on forced labor, and to work on implementing Paris Agreement on climate.

The agreement prohibits the renationalization of previously liberalized sectors, and China has allowed market access in many sectors, including telecoms, where China has lifted the ban on cloud services. In this sector, European investors can acquire 50 percent of the company's shares. According to the deal, China will also implement "technology neutrality" clause, which means the above-mentioned equity cap on telecommunications sectors will not be implemented in the case of financial, logistical, and medical services of offered one-line (European Commission, December 30, 2020).

That Brussels and Beijing were able to make these necessary compromises and agree on the main points of the agreement must have come as a surprise to many analysts after seven years of negotiations. For example, just a week before the agreement was reached, Mears and Leali published an article in which they collected the opposing voices on the deal and concluded that the chances of reaching an agreement with China were diminishing (Mears & Leali, December 22, 2020).

The president of EU Chamber of Commerce in China expressed his doubts in the summer of 2020 that the deal could be completed in time (Reuters, June 23, 2020). After the political deal was struck, several analysts were critical of whether it was worth finalizing the agreement with Beijing. Andreas Kluth underlines the long-term political significance of the deal and calls it a mistake in his analysis:

> Thinking small, the Europeans seem to have welcomed Xi's overture. Realizing that Beijing is in a hurry before Biden's inauguration, they tactically pocketed a few token concessions by China—still to be clarified—and proclaimed success. In doing so, they may have jeopardized what should be their bigger strategic goal: a united Western front to compel China to genuinely accept a liberal and rules-based international economic model.
>
> (Kluth, 2020)

Strengthening the liberal international model, as we have already underlined, is not a realistic foreign policy goal when making an investment deal.

The similarly critical but more realistic critic points out how American foreign policy has lost a long march against China. Mitchel and Manson have
described the last leg of that march:

> In the end, it was China's president Xi Jinping who would steal a march
> on his US rival by signing both the CAI and the Regional Compre-
> hensive Economic Partnership, a separate regional deal with many of
> America's closest Asia-Pacific allies, in the waning days of Mr. Trump's
> administration.
>
> (Mitchel & Manson, 2020)

Opponents of the deal stress that what will matter is the details of the deal
and practice on the ground. Several MEPs want to fight over labor standards and human rights, arguing that the trade deal is related to how (alleged)
forced labor is dealt with (Mears & Leali, December 22, 2020).

In conclusion, the deal has been made but the fight is not over yet, and
until the agreement can also be signed, there are still many months to go.
To make the situation even more confusing, the United States has very clear
foreign policy goals in pushing EU for a tougher stance on China, which
efforts have been partially rewarded, but in this case, the question remains
whether the EU has the means to oppose the United States and whether it
can pursue its own goals. As Borell put it:

> Amid US-China tensions as the main axis of global politics, the pres-
> sure to 'choose sides' is increasing, … We as Europeans have to do it
> 'My Way', with all the challenges this brings.
>
> (Borell, 2020)

The next subsection raises the question of what the regulatory framework looks like at the EU level and how the EU's competitive position has
changed in recent years with regard to technological development and 5G
in particular.

2.3 Technological development and regulatory framework in the EU

2.3.1 *Technological development and its policies in the EU*

There is no doubt that the regulation and the national legal frameworks have
the potential to significantly impact Chinese investment in the Single Market. The likelihood of significant impact increases when direct investment
targets technology companies that are leaders in technology development.
This new approach, which implies more protectionist trade and investment
policies than ever before, has long-term policy implications. Until now, the
EU has been the embodiment of the free trade policy approach and has also

welcomed foreign investment. This is still the case; however, the EU has recently begun to tighten the rules, partly in response to similar practices in the United States, Russia, and China. As we understand it, the second wave of globalization may have already peaked with the Global Financial Crisis and was put on hold with the devastating effects of the Covid-19 global pandemic on trade and investment freedom. Although the new European approach to trade, investment, and technology transfer is logical and easy to argue for at the local level—given the reactions of its main competitors—it is self-defeating at the global level if all actors behave in the same way.

We do not think that the EU would need more protectionist tools but more funding to accelerate technological development. The EU generally seems to lag behind its main competitors in R&D spending. The EU spent 2.19 percent of its GDP on R&D in 2019, compared to 2.82 percent in the United States, 3.28 percent in Japan, and 4.52 percent in Korea in 2018. China's R&D intensity was about the same (2.06 percent in 2018). The amount spent on R&D in the EU was $428 billion in 2018, compared to $462 billion in China and $551 billion in the United States, according to the World Bank.

At the same time, a comparison of corporate spending on R&D shows that European companies' position is not that bad. According to 2020 EU Industrial R&D Investment Scoreboard, corporate R&D spending in the EU was $189 billion, while the Chinese corporate sector spent $119 billion and American firms $349 billion on R&D.[9] As the number of research-intensive companies is concerned, Europe lags behind the main competitors. The United States has 775 and China 536 companies in the comprehensive database of 2,500 companies worldwide that invest significant resources in R&D, while EU member states only have 421 firms on the list (European Commission, 2020c). Looking at the list, only four German firms make it to the top 20 of these firms, and all of them are from the automotive industry.

The conclusion which can be drawn from these figures and trend is that we have a dangerous cocktail of factors that further worsen EU opportunities, the relatively low R&D spending in general and at corporate level compared to the United States, Japan, and China, and trends toward restricting capital and technology inflows.

According to the EC, the overall cost of 5G network and full fiber infrastructure in the EU is around €500 billion, while the EU budget for R&D is not significant from this point of view. The foreseen amount for R&D spending is €95.5 billion between 2021 and 2027. It is not clear exactly how much of this amount will be spent on 5G research and innovation; but in hindsight we can see that this amount from the R&D budget between 2014 and 2020 (Horizon 2020 program) was a meager €700 million, which does not bode well for the future (European Commission, 2014: 2). It is obvious that not all the funding must come from the public sphere, but even (1) corporate financing is weak due to the size of market, (2) public funding is weak too due to unsolved institutional problem, and (3) the weakness of innovation.

2.3.1.1 Weak corporate financing

The EC also made it clear that it believes the two European tech firms, Nokia and Ericsson, are capable of providing EU complete 5G environments. Only Huawei could provide complete solutions, while Samsung and China's ZTE can only partially replace Huawei and the other two European firms (Bellamy, 2020). The question is what is more important for European policymakers: fostering European champions and focusing on the long-term goal of "technological sovereignty" or accelerating the establishment of 5G networks in Europe and maintaining competitiveness, although the goals are more short and medium term.

Solomon points out that the EU one-time world leader in mobile phones and networks is now at a competitive disadvantage. He lists indicators such as the percentage of 4G in total subscriptions, mobile download speeds, and household fiber penetration, and shows how the EU and selected countries are lagging behind (Solomon, 2020, July 1).

We must add that lagging can be explained by the relatively small markets in the EU. The fragmented markets make it difficult for the already debt-ridden industry to invest heavily in 5G development. According to Bloomberg data, the value of European telecom companies nearly halved between 2012 and 2018 (2012: $234 billion; 2018: $133 billion), while the value of American companies increased by 71 percent (2018: $532 billion) and the value of Asian companies increased by 13 percent (2018: $561 billion) (Fildes, 2019). The EC cites a European Investment Bank assessment of the financing of 5G networks:

> There is consensus among experts that market forces will not guarantee the achievement of the Digital Agenda for Europe and European Gigabyte Societies targets. According to a recent study commissioned by the EIB (forthcoming), the estimated investment needs to meet such targets as from 2018 amount to €345-360bn for the EU27 (€380-395bn for the EU28). Expected private funding will cover about one third of this amount, leaving an estimated investment gap on an annual basis of around 42bn€ until 2025. As the private funding baseline was projected before the COVID crisis, the gap may have increased due to investment cut backs in the private sector...
>
> (European Commission, 2020d)

2.3.1.2 Weak public funding

We have already stressed that an effective industrial policy could theoretically compensate for this weakness of the European market, but industrial policy raises the question of whether national champions should be promoted at EU level or at member state level. Besides this question, we must face another dilemma too, that of the unfunded two-thirds funding, which needs urgent resolution solution. There are two reasonable options: creating more competition and relaxing regulations to attract more capital; let's call it a *market-friendly approach*. The second option is to spend significantly higher amounts of earmarked EU or national funds to close the funding gap; let's call it an *interventionist approach*. Both solutions have their downsides:

1   The *market-friendly approach* is less suited to ensure security needs in an increasingly (geo)politicized business environment. The EU-China Comprehensive Investment Agreement might be the first signal over the past years that the EU aims to achieve technological sovereignty in practice too and it does not blindly follow the guidance of the American foreign policy.
2   The *interventionist approach* can mitigate security concerns and help to promote national champions, but it is contrary to the logic of the Single Market or the EU if this logic is applied at Member State level, and it would contradict the concept of sovereign states forming the EU; moreover, if logic of efficiency (better firms get more support regardless of their origin) would get the upper hand, even this would run into the opposition as Germany—based on their market position—would get the most support, as Germany alone has 212 companies on the list of the 1,000 most research-intensive EU and UK firms. The second EU country is France, where corporate R&D spending was €35 billion and 122 firms on the list.

Based on the political reactions, neither a purely interventionist approach at EU level nor a market-friendly approach will be supported; but a mixture of both approaches could help to "elevate" European champions, striking a balance between efficiency and safeguarding national interests.

2.3.1.3 The weakness in innovation and 5G sector

The 2020 EU Industrial R&D Investment Scoreboard compares the EU's position in four areas of corporate R&D expenditure (health, automotive, information and communications technology [ICT] producers, and ICT services) with the United States and China. Comparing the EU and the United States, the report concludes that EU investment in R&D is led by traditional mid-tech sectors (automotive), while the United States focuses on high-tech sectors (ICT). Comparing the EU and China, the EU dominated in all sectors in 2010, while the lead in ICT sectors was taken by Chinese companies in 2019 (European Commission, 2020c: 53).

What is more important for the 5G sectors is patent activity, which signals future trends well. The two major European tech companies lag behind their competitors in 5G patents. Nokia owns 12 percent of the core patents (Standard Essential Patent), while Ericsson owns 9 percent of these core patents. Huawei's share is 19 percent of core patents, while Korea's Samsung owns 15 percent (LG: 14 percent; Qualcomm: 14 percent). A study by Tim Pohlman, founder of commercial technology consultancy IPlytics, was officially commissioned by the German Federal Ministry for Economic Affairs and Energy, findings were summarized thus:

> The results of the study show that more and more 5G patent owners are coming from China. The Chinese technology provider Huawei declared most families for 5G and registered them internationally in all countries. The statistics on standard contributions support Huawei's strong position in the development of the 5G standard. With expenditures of over 15 billion US dollars in research and development (R&D) in 2018, which according to Huawei's management were primarily invested in the further 5G development, Huawei is among those companies spending most dollars on 5G related R&D. However, the study also shows that companies such as Nokia, Ericsson and Qualcomm, which were leaders in previous 2G, 3G, and 4G generations, are also playing a leading role in developing the 5G standard.
>
> (Pohlman, 2020: 42)

Despite these conclusions, we can argue that the trade dispute between the United States and China and bans on the Huawei distorted market positions and opened up new opportunities for firms such as Ericsson and Nokia in 5G infrastructure.

2.3.2 The regulatory framework

The EU does not have a common 5G or AI policy; the responsibility for these areas lies with the member states. However, the EC is making significant efforts to coordinate the different approaches to solving problems or issues in these areas. Following the European Council's call for the EC to work together to address the issues, the EC has issued a recommendation on cybersecurity of 5G networks in 2019 (European Commission, 2019b).

Based on this recommendation, member states carried out national risk assessments and reviews of national measures. Based on these reports, the Network and Information Security (NIS) cooperation group published a report (NIS Cooperation Group, 2019) which, together with the European Union Agency for Cybersecurity landscape mapping—an annual assessment on emerging threats—formed the basis for the 5G toolbox (European Commission, 2020b). Table 2.4 shows the timeline of how regulation and policy regarding 5G started to emerge at the EU level.

In mid-2020, the first report on 5G toolbox implementations was produced and released by the NIS collaborative group (NIS Cooperation Group, 2020). The main conclusion of the report was that good progress has been made in strengthening the powers of national authorities by limiting the involvement of suppliers based on their risk profile. At the same time,

Table 2.4 EU and member states' actions in the field of 5G regulation

Player	Content	Date
EC	Action plan	September 2016
European Council	Conclusions call on adopting a recommendation on cybersecurity of 5G networks	March 2019
EC	Recommendation on cybersecurity of 5G networks	March 2019
Member States	National Risk Assessments	July 2019
Member States	Report on EU coordinated risk assessment of 5G networks security	October 2019
EC	EU toolbox for the Member States that helps to mitigate the cybersecurity risks	January 2020
NIS Cooperation Group	Report on Member States' progress in implementing the EU toolbox on 5G cybersecurity	July 2020
EC	Commission recommendation on a common Union toolbox for reducing the cost of deploying very high capacity networks and ensuring timely and investment-friendly access to 5G radio spectrum	September 2020

Source: Own compilation.

according to the report, the risk of dependency on "high risk" suppliers needs to be mitigated. Mitigating risk dependency is one of the key words in the report. The report also points out that FDI screening is critical to ensuring security. This is the reason why we focus so much on FDI screening mechanisms later in the case studies. The report also adds that the 13 countries in the EU-27 that have not yet adapted and implemented FDI screening mechanisms need to put these regulations in place. The text here refers to EU's FDI screening regulation, which was passed in 2018 and came into force in April 2019 (European Commission, 2019c). The regulation created a new coordination mechanism where the EC and member states can share information and air concerns about specific investments as needed. The EU has taken a liberal policy approach (compared to other OECD countries) in establishing the screening mechanism, which is more of a platform for EU countries to work together. Since the implementation of the framework, extensive comparative research has been conducted on national schemes (see more on this in Chapters 4 and 5).

In this chapter we have shown that at the heart of the political and economic debates between China and the EU is the European expectation that China's economic progress will also become social progress, and that the

EU—because of its values, principles, and economic influence in China—can accelerate this progress.

We have also pointed out that China also has a growing but relatively weak economic influence in the EU. However, this weak, rising influence is often misinterpreted and exaggerated in the literature and media. We argue that there is essentially one segment where the added value from cooperation with China would be very beneficial for the EU, namely the 5G segment, but it is this segment that is heavily distorted by government interventions as it causes the most security and foreign policy concerns for the EU and member states.

At the same time, the first chapter of this book helped us understand why negotiating partners sometimes seem to be in "parallel universes" and why communication and cooperation with China becomes much easier to understand if we interpret China as a developmental state. This paradigm can also provide clues to understanding what can and cannot be changed about China's behavior. This is also why the political deal on the EU-China CAI might have come as a shock to the United States, while the United States was able to convince the majority of EU member states to ban Chinese firms from the 5G market. Huotari claims that the details of the deal are not yet known; he assumes that the deals could also include changes in the European attitude toward China in this area. He warns:

> If Europe concludes this agreement, it is likely to lose some trust with like-minded partners including the U.S. who do not necessarily see this deal as a sign of Europe's "strategic autonomy."
>
> (Huotari & Zenglein, 2020)

We disagree with this statement because, as we understand it, this European behavior is precisely the essence of strategic autonomy, to make decisions independently, taking into account our interests. In the next chapter, we focus on the evolution of EU and United States relations in the light of China cooperation and on two key European countries—Germany and Russia—in this political game. German interests are crucial in shaping the EU's room for maneuver, and Russian interests are essential to consider if we are to understand what options China has in seeking an ally or partner in order to reshape the global order.

Notes

1 Central European economies were transformed in the early 1990s, and the transformation was based mainly on capital and technology from the West, and to this day, these Western European firms dominate these markets. See more on transformation in Salamon (1995). It also means that Western European companies with significant interests in the Central and Eastern European region would lose from the growing Chinese economic presence in that region.

2 In the case of China, it is more difficult for the United States to ally interests and strategies when countries outside EU are involved. The 17+1 cooperation framework is a good example where the interests of the EU member states and non-EU member states can be easily distinguished and juxtaposed. EU member states that have access to EU transfers (i.e., grants) have greater leeway in choosing appropriate funding instruments, while non-EU members are more restricted in their choices.

3 The deal between the EU and China will have implications for the US-China negotiations and it would most likely harden the American position. Delany and Klein conclude that:

> No one in the United States will feel the pressure more than President Donald Trump, who has staked much of his reputation on creating a more balanced trade and investment relationship with China. Any concessions the EU might extract from Beijing would likely intensify the American leader's hard-driving approach to a bilateral trade war with China that has dragged on for two years, analysts said.
>
> (Delany & Klein, 2020, July 26)

The temptation for the United States to ally with the EU against China is strong, and the European reaction to this American approach has been neither clear nor enthusiastic, but there is already cooperation between the United States and the EU on China. One example is the WTO negotiations on industrial subsidies. The EU, Japan, and the United States issued a joint statement in early 2020 agreeing that new types of unconditionally prohibited subsidies should be added to the prohibitive list, namely:

> a. unlimited guarantees; b. subsidies to an insolvent or ailing enterprise in the absence of a credible restructuring plan; c. subsidies to enterprises unable to obtain long-term financing or investment from independent commercial sources operating in sectors or industries in overcapacity; d. certain direct forgiveness of debt.
>
> (Joint Statement of the Trilateral Meeting of the
> Trade Ministers of Japan, the US and the EU)

4 Germany ranked 9th, France 13th, and Italy 36th in the same ranking in 2016.

5 Spross argues that if a poor country wants to develop, it must open for advanced countries. While doing so, it offers natural resources or cheap labor and it most often ends up buying high value-added content from the advanced countries. However, if these countries import high value-added content and export low value-added content, they often face trade deficit. The advised recipe to solve the problems is often austerity and tight monetary policy which leads to new waves of deregulation to attract more foreign capital. The only way to break out from this cycle is to set condition on the investors in order to accelerate the transfer of skill and even technology (Spross, 2019).

6 According to the consultation summary, several member countries pointed out that there are difficulties in collecting information on foreign subsidies, and it also mentions possible negative effects on administrative burden and foreign investment. Countries also stress that the new instrument regulating foreign state aid should not be stricter than the existing EU state aid rules.

7 It should be added that the data come from the EU's Foreign Ownership Database, which does not include firms with fewer than ten employees. The simple reason is that these firms often do not produce balance sheets or other documents necessary to identify ownership (European Commission, 2019d: 11).

8 Attitudes toward Chinese direct investment and the regulatory environment have been tightened in recent years in the United Kingdom, but the country is still attractive to Chinese investment. However, it is not clear that the uncertainty caused by Brexit and the Huawei ban will not create a very different environment in a few years. Given the traditionally strong ties between the United States and the United Kingdom, it was not surprising that London banned Huawei from the country's telecommunications networks, including 5G and earlier generations, by the end of 2027 (Moldicz, 2020: 3–36).

9 Japan $115 billion on R&D in 2019.

References

Bellamy, D. (2020, June 25). EU Insists European Companies Could Replace Huawei in 5G Network. *Euronews*, Retrieved from: https://www.euronews.com/2020/07/25/eu-insists-european-companies-could-replace-huawei-in-5g-network

Bergsten, F. C. (1982, Summer). What to Do about the U.S. Japan Economic Conflict. *Foreign Affairs*, pp. 1059–1075. Retrieved from: https://www.foreignaffairs.com/articles/japan/1982-06-01/what-do-about-us-japan-economic-conflict

Borell, J. (2020, June 09). EU-China Strategic Dialogue: Remarks by High Representative/Vice-President Josep Borrell at the Press Conference, Retrieved from: https://eeas.europa.eu/headquarters/headquarters-homepage/80639/eu-china-strategic-dialogue-remarks-high-representativevice-president-josep-borrell-press_en

Council of the European Union (2003). European Security Strategy. A Secure Europe in a Better World, Retrieved from: https://www.consilium.europa.eu/en/documents-publications/publications/european-security-strategy-secure-europe-better-world/

Delany, R. & Klein, J. X. (2020, July 26). US Trade Negotiators May 'Smell Blood in the Water' If China Makes Concessions in EU Talks. *South China Morgan Post*, Retrieved from: https://www.scmp.com/economy/china-economy/article/3094684/us-trade-negotiators-may-smell-blood-water-if-china-makes

European Commission (2014, February 11). 5G Infrastructure Public Private Partnership (PPP): The Next Generation of Communication Networks Will Be "Made in EU." *Factsheet/Infographic*, Retrieved from: https://ec.europa.eu/digital-single-market/en/news/5g-infrastructure-public-private-partnership-ppp-next-generation-communication-networks-will-be

European Commission (2019a). EU-China. A Strategic Outlook, p. 1, Retrieved from: https://ec.europa.eu/commission/sites/beta-political/files/communication-eu-china-a-strategic-outlook.pdf

European Commission (2019b). Commission Recommendation of 26.3.2019. Cybersecurity of 5G Networks. *Strasbourg*, Retrieved from: https://ec.europa.eu/digital-single-market/en/news/cybersecurity-5g-networks

European Commission (2019c). Regulation (EU) 2019/452 of the European Parliament and of the Council of 19 March 2019 Establishing a Framework for the Screening of Foreign Direct Investments into the Union, Retrieved from: https://eur-lex.europa.eu/legal-content/EN/TXT/PDF/?uri=CELEX:32019R0452&from=EN

European Commission (2019d). Commission Staff Working Document in Foreign Direct Investment in the EU. Following up on the Commission Communication "Welcoming Foreign Direct Investment While Protecting Essential Interests" of

13 September 2017, Retrieved from: https://trade.ec.europa.eu/doclib/docs/2019/march/tradoc_157724.pdf

European Commission (2020, December 30). Key Elements of the EU-China Comprehensive Agreement on Investment. Press Release, Retrieved from: https://ec.europa.eu/commission/presscorner/detail/en/ip_20_2542

European Commission (2020, January 29). Secure 5G Networks: Commission Endorses EU Toolbox and Sets Out Next Steps. Press Release, Retrieved from: https://ec.europa.eu/commission/presscorner/detail/en/ip_20_123

European Commission (2020, June 22). EU-China Summit: Defending EU Interests and Values in a Complex and Vital Partnership. Brussels. Press Release, Retrieved from: https://www.consilium.europa.eu/hu/press/press-releases/2020/06/22/eu-china-summit-defending-eu-interests-and-values-in-a-complex-and-vital-partnership/

European Commission (2020a). White Paper on Levelling the Playing Field as Regards Foreign Subsidies, Retrieved from: https://ec.europa.eu/competition/international/overview/foreign_subsidies_white_paper.pdf

European Commission (2020b). Cybersecurity of 5G Networks. EU Toolbox of Risk Mitigating Measures, Retrieved from: https://ec.europa.eu/digital-single-market/en/news/cybersecurity-5g-networks-eu-toolbox-risk-mitigating-measures

European Commission (2020c). EU R&D Scoreboard. The 2020 EU Industrial R&D Investment Scoreboard, Retrieved from: https://iri.jrc.ec.europa.eu/scoreboard/2020-eu-industrial-rd-investment-scoreboard

European Commission (2020d). Identifying Europe's Recovery Needs. Accompanying the Document Commission Staff Working Document. Communication from the Commission to the European Parliament, the European Council, the Council, the European Economic and Social Committee and the Committee of the Regions. Europe's Moment: Repair and Prepare for the Next Generation, Retrieved from: https://eur-lex.europa.eu/legal-content/EN/TXT/PDF/?uri=CELEX:52020SC0098&from=EN

European Council (2020, June 22). EU-China Summit via Video Conference, Retrieved from: https://www.consilium.europa.eu/en/meetings/international-summit/2020/06/22/

European Parliament (2019). European Parliament Resolution of 12 March 2019 on Security Threats Connected with the Rising Chinese Technological Presence in the EU and Possible Action on the EU Level to Reduce Them. 2019/2575(RSP), Retrieved from: https://www.europarl.europa.eu/doceo/document/TA-8-2019-0156_EN.html

European Union Chamber of Commerce (2020). European Business in China. Business Confidence Survey 2020. *Navigating in the Dark.*

Fildes, N. (2019, January 27). 5G: Can Europe Match the US and China on Mobile Networks? *Financial Times*, Retrieved from: https://www.ft.com/content/650d3bf8-1e32-11e9-b2f7-97e4dbd3580d

Guluzade, A. (2019, May 7). The Role of China's State-Owned Companies Explained. *World Economy Forum*. Blogpost, Retrieved from: https://www.weforum.org/agenda/2019/05/why-chinas-state-owned-companies-still-have-a-key-role-to-play/

Huotari, M. & Zenglein, M. J. (2020, December 22). The EU-China Comprehensive Agreement on Investment (CAI) is a Test for the Future Trajectory of the EU-China Relationship. *MERICS*, Interview, Retrieved from: https://merics.

org/en/interview/eu-china-comprehensive-agreement-investment-cai-test-future-trajectory-eu-china

Joint Statement of the Trilateral Meeting of the Trade Ministers of Japan, the United States and the European Union (2020, January 14). Washington, Retrieved from: https://trade.ec.europa.eu/doclib/docs/2020/january/tradoc_158567.pdf

Kluth, A. (2020, December 30). The China-EU Investment Deal Is a Mistake. *Bloomberg*, Retrieved from: https://www.bloomberg.com/opinion/articles/2020-12-30/europes-big-investment-deal-with-china-is-a-mistake

Kratz, A., Huotari, M., Hanemann, T. & Arcesat, R. (2020, April 08). Chinese FDI in Europe: 2019 Update. *Rhodium Group (RHG) and MERICS*, Retrieved from: https://mimderics.org/en/report/chinese-fdi-europe-2019-update

Mead, W. R. (2021, January/February). The End of the Wilsonian Era. Why Liberal Internationalism Failed. *Foreign Affairs*, Retrieved from: https://www.foreignaffairs.com/articles/united-states/2020-12-08/end-wilsonian-era

Mears, E. & Leali, G. (2020, December 22). EU-China Investment Deal Hits a Snag as US Exerts Pressure. *Politico*, Retrieved from: https://www.politico.eu/article/eu-china-investment-deal-no-show-us-forced-labor/

Mitchel, T. & Manson, K. (2021, January 1). China Sees EU Investment Deal as Diplomatic Coup after US Battles. *Financial Times*, Retrieved from: https://www.ft.com/content/64ef5592-25b4-48c4-a70b-b42071951941

Moldicz, Cs. (2020). Chinese Direct Investments in the EU and the Changing Political and Legal Frameworks. *Contemporary Chinese Political Economy and Strategic Relations*, 6(1), pp. 3–36.

NIS Cooperation Group (2019, October 9). EU-Wide Coordinated Risk Assessment of the Cybersecurity of 5G Networks. Report, Retrieved from: https://ec.europa.eu/digital-single-market/en/news/eu-wide-coordinated-risk-assessment-5g-networks-security

NIS Cooperation Group (2020, July 24). Report on Member States' Progress in Implementing the EU Toolbox on 5G Cybersecurity, Retrieved from: https://ec.europa.eu/digital-single-market/en/news/report-member-states-progress-implementing-eu-toolbox-5g-cybersecurity

Peng, M. W., Ahlstrom, D., Carraher, S. M. & Shi, S. W. (2017, March). History and the Debate over Intellectual Property. *Management and Organization Review, 13*(1), pp. 15–38. Retrieved from: https://personal.utdallas.edu/~mikepeng/documents/Peng17_MOR_A%20C%20S_13(1_March)15_38.pdf

Pohlmann, T. (2020, January). Fact Finding Study on Patents Declared to the 5G Standard. *IPlytics*, Retrieved from: https://ipforbusiness.org/wp-content/uploads/2020/02/5G-patent-study_TU-Berlin_IPlytics-2020.pdf

Reuters (2020, June 23). EU Business Body Fears China-EU Investment Deal Will Not Be Finished This Year. *Reuters*, Retrieved from: https://www.reuters.com/article/us-china-eu-investment-idUSKBN23U0WI

Salamon, R. (1995). *The Transformation of the World Economy*. New York: St. Martin's Press, p. 238.

Scissors, D. (2019, January). Chinese Investment: State-Owned Enterprises Stop Globalizing, for the Moment. *American Enterprise Institute*, Retrieved from: https://www.aei.org/wp-content/uploads/2019/01/China-Tracker-January-2019.pdf?x88519

Small, A. (2019, April 3). Why Europe is Getting Tough on China? And What It Means for Washington. *Foreign Affairs*, Retrieved from: https://www.foreignaffairs.com/articles/china/2019-04-03/why-europe-getting-tough-china

Solomon, G. (2020, July 1). The State of European Connectivity. How Ready Are We for 5G? *Ericsson Blog*, Retrieved from: https://www.ericsson.com/en/blog/2020/7/the-state-of-european-connectivity-how-ready-are-we-for-5g

Spross, J. (2019, April 1). China's Forced Technology Transfer Is Actually a Pretty Good Idea. *The Week*, Retrieved from: https://theweek.com/articles/831859/chinas-forced-technology-transfer-actually-pretty-good-idea

World Economic Forum (2016). Global Enabling Trade Report 2016. *World Economic Forum and the Global Alliance for Trade Facilitation*. Insight Riport, Retrieved from: http://www3.weforum.org/docs/WEF_GETR_2016_report.pdf

3 Economic and political interests of the major powers
The United States, Germany, and Russia

3.1 The changing American stance on China

Aside from the question of how China's advances in modern technology can transform global order, this subsection focuses on American economic interests which are less articulated in this debate, as foreign policy issues are at the forefront of the discourse. In our view, the geopolitical and geoeconomic approaches together can help us understand American motivations for reshoring, technological decoupling, guiding European foreign policy, and so on. This is also why the EU and the US positions on trade with China are compared to show differences that lead to different positions toward China. What needs to be understood is that the United States as a hegemonic power, is more concerned by political factors than economic ones, while the EU seems to prefer the enforcement of business interests.

3.1.1 From the "pivot to Asia" concept to the "new cold war"

After decades of a policy of engagement with China, the first signs of change appeared on the radar in the 2010s when Hillary Clinton's article "America's Pacific Century" was published in *Foreign Affairs*. The article signaled a definite shift in American foreign policy in 2011. She wrote:

> The future of politics will be decided in Asia, not Afghanistan or Iraq, and the United States will be right at the center of the action.
>
> (Clinton, 2011)

The recognition that the gravitational center of the world economy and politics was shifting to Asia, and China in particular, had been there for decades; however, the rise of China has never been more evident. The turn to Asia also meant that Europe was less in focus after decades of Cold War tensions on that continent.

As we understand it, the recent American efforts to push back on China need to be interpreted in a broader perspective than just China-EU relations, as the Americans are simultaneously trying to convince their European

DOI: 10.4324/9781003128625-3

allies to take a tougher stance on Russia as well. The two-pronged foreign policy approach is motivated by Chinese technological advances in 5G, artificial intelligence (AI), and other related areas that threaten American superiority and Russian military development, along with a more assertive foreign policy that is causing headaches for American policymakers.

As tensions increased in US-China relations, the term "new cold war" was coined. Although the phrase was often used in the media to highlight the similarities between the recent trade war between China and the United States and the contest between the North Atlantic Treaty Organization (NATO) and Warsaw Pact countries, the analogy is deeply flawed for several reasons:

1   Our world economy is much more interconnected today than it was in the 1950s because of the immensely intertwined global supply chains. There are clear American efforts to bring manufacturing home from China; however, this will take at least a decade as supply chains (clusters of firms, the corporate culture, etc.) took more than two decades to create. The question is whether the usually more short-lived political interests will prevail over the long-term economic interests.

2   Not only the world economy, but the American and Chinese economies are now much more intertwined than trade relations between the Soviet Union and the United States ever were. Ferguson and Schularick used the term "Chimerica" to describe the symbiotic relationship between the United States and China in 2006; but given the changing conditions, they also announced the end of "Chimerica" in 2018. Although the "trade war" had a significant impact on trade, China accounted for 18.1 percent of US merchandise trade and 6 percent of US merchandise exports in 2019. The stock of US foreign direct investment (FDI) in China was $116.2 billion in 2019 and the stock of China's FDI in the United States was $37.7 billion in 2019, according to Office of the United States Representative.

3   While striving for greater influence in the world, China shows no signs of spreading its economic and political model, unlike the way the Soviet Union did, spending resources not only in Eastern Europe, but also in Africa, Asia, and South America to promote the ideology of Leninism and Marxism and its political and economic model. It must be stressed that exporting the economic and political model is different from being a Leninist or Marxist party. In addition, the Chinese Communist Party (CCP) has abandoned much of the original ideas (class struggle, labor theory of value, socialization of the means of production, etc.) and focuses more on maintaining power. At the same time, it is also clear that China is actively engaged in exporting elements of its model dedicated to state control over society in African countries (Economy, 2020: 4).

4   The Soviet Union has never seriously challenged the United States in terms of economic power, but China has. According to the International

Monetary Fund, China's GDP in 2020 was 75 percent of the GDP of the United States in current prices, while the Soviet Union's GDP before its collapse was less than one-fifth. At the same time, China's military power (especially nuclear weapons) is inferior to that of the United States.

There is only one item in this list that can be achieved by consistent political steps, and that is at least partial decoupling from China. In addition to raising tariffs and limiting Chinese access to the global financial infrastructure, decoupling policies have two elements that more directly affect technology competition between China and the United States: the reshoring of global supply chains and technological decoupling.

3.1.1.1 Reshoring

The rise of global supply chains between 1990 and 2019 led to a worldwide concentration of manufacturing output. In 2018, 28.4 percent of global manufacturing output came from China, which proved to be a geopolitical risk during the global pandemics, especially during the first wave which affected the flow of goods immediately and exposed the vulnerability of Western countries. Table 3.1 shows the share of the top ten countries in global manufacturing output. In some sectors, dependence on China is extremely high; for example, in the pharmaceutical sector, about 40 percent of active ingredients come from China (Rotondi et al., 2020).

Recently, in the aftermath of the Covid-19 pandemic, the US Department of Commerce has launched initiatives to attract American firms from China and provide them with tax incentives and government subsidies (Pamuk & Shalal, 2020). These efforts are not new, but the Trump administration has made its goal—"bring manufacturing home" from China—very explicit (Donnan & Hamlin, 2018).

Table 3.1 Top ten countries' share in global manufacturing output in 2018 (added value, in current prices, percentage)

China	28.4
US	16.6
Japan	7.2
Germany	5.8
South Korea	3.3
India	3.0
Italy	2.2
France	1.9
United Kingdom	1.8
Mexico	1.5

Source: Richter (February 25, 2020).

Table 3.2 US manufacturing import ratio (import of 14 Asian countries[a] as of domestic manufacturing gross output, percentage)

Year	2008	2009	2010	2011	2012	2013	2014	2015	2016	2017	2018	2019
Ratio	9.2	9.5	10.5	10.3	10.6	10.6	11.2	12.3	12.5	12.7	13.1	12.1

Source: Van den Bossche et al. (2020).
a The 14 Asian countries are China, Vietnam, Philippines, Malaysia, Indonesia, Pakistan, Sri Lanka, Taiwan, Thailand, Bangladesh, India, Singapore, and Hong Kong.

Table 3.3 Chinese share in the total export of 14 Asian countries to the United States (percentage)

2018Q1	2018Q2	2018Q3	2018Q4	2019Q1	2019Q2	2019Q3	2019Q4
67	65	66	65	60	61	59	56

Source: Van den Bossche et al. (2020).
a The 14 Asian countries are China, Vietnam, Philippines, Malaysia, Indonesia, Pakistan, Sri Lanka, Taiwan, Thailand, Bangladesh, India, Singapore, and Hong Kong.

The mixture of steps proved successful, as American manufacturing required less input from Asia and the share of Chinese imports of manufactured goods also declined significantly. The decline in the Asian share of imports in manufacturing is much less significant than the Chinese share, leading us to conclude that Chinese exports are being replaced by other Asian countries, especially Vietnam. Table 3.2 shows the Asian import share in US manufacturing.

In 2019, Chinese imports of manufactured goods fell 17 percent, or $90 billion, from the previous year. In the same year, imports from 13 other Asian countries increased by $31 billion. Add Mexico, which increased its exports to the United States by $13 billion, and the assessment may conclude that the United States was able to divert trade away from China to some degree. Additional years can be seen in Table 3.3, which shows that the Chinese share of total exports of the 14 Asian countries changed from the beginning of 2018 to the end of 2019. Despite the seemingly compelling data, we should not forget that there are strong incentives to transship Chinese manufactured goods through Vietnam to reach American markets. The sudden surge in Chinese-Vietnamese trade suggests the likelihood of this scenario.

3.1.1.2 The consequences of technological decoupling

The "new cold war" is not just about trade barriers and exchange rates, but it is a technological competition between the United States and China, in areas like AI and 5G. The current tone of American political discourse is best captured in a quote from Eric Schmidt, the former chairman of Google, who famously spoke about the Chinese. Mehta quotes him:

> By 2020, they will have caught up. By 2025, they will be better than us. And by 2030, they will dominate the industries of AI.
>
> (Mehta, 2017)

In arguing why the United States should be afraid of China's technological superiority, two elements are highlighted: (1) reshaping cyberspace along Chinese lines would make the Internet less global and less open; (2) in this case, the benefits of technological dominance would be reaped by China, not the United States (Segal, 2018).

1   *Remaking cyberspace.* This battle for control of cyberspace did not start with the banning of Huawei, TikTok, or WeChat, but with the banning of Google services in China. The reasoning is different, but the purpose is the same: to dominate cyberspace by displacing competitors and limiting ideological influence. The long-term effect of forced technological decoupling is the creation of (at least) two technological microcosms or ecosystems. The story revolving around the Huawei Corporation (TikTok, WeChat) may be the first harbinger of this new world, a world where you have to take sides when you buy your phone, choose applications, and use software. The American promise of a free internet has been deeply discredited in recent years, and while the Chinese concept of cyber sovereignty may sometime be attractive to some nations who wish to preserve their sovereignty rather than promote a US-dominated technological environment. Some scholars emphasize that the United States would "distribute" and thus delegate authority to several network regulatory bodies. Segal adheres to this view; he puts it this way:

> Washington and its allies have promoted a distributed model of Internet governance that involves technical bodies, the private sector, civil society, and governments, whereas Beijing prefers a state-centric vision.
>
> (Segal, 2018)

What must be clear to us is that delegation of various tasks does not necessarily lead to weaker control. Whistleblower Edward Snowden summarized the current American approach to surveillance in an interview with Heuvel and Cohen in *The Nation*:

> Richard Nixon got kicked out of Washington for tapping one hotel suite. Today we're tapping every American citizen in the country, and no one has been put on trial for it or even investigated. We don't even have an inquiry into it.
>
> (Snowden cited by Vanden Heuvel & Cohen, 2014)

So, it is not surprising that China has recently made considerable efforts to strengthen its technological independence, which more or less probably leads to two separate ecosystems, or at least to a world in which (technological) permeability decreases. The irony is that the American decoupling policy only accelerates this process, making China more independent and less connected to the world, while the EU and its member states will have to take sides unless they manage to strengthen their technological sovereignty.

2   *The benefits of technological supremacy.* This argument brings us closest to answering the question of why the United States adopted the strategy of technological decoupling toward China. The United States has acted as a hegemonic power since World War II (Edelstein & Krebs, 2015: 110–111) and understandably, as a hegemon, the United States seeks to protect its positions while China struggles to gain more power and influence in the global economy and politics. The behavior of both China and the United States is motivated by the pursuit of foreign policy interests, but at the same time they are the result of longstanding features of the domestic and international field in both countries. In the United States, corporate interests, wealth, and certain elements of American political culture are the driving forces behind the American "mission awareness," while the characteristics of Chinese political institutions, historical development (bad experiences with the West), and the interests of global Chinese firms have a major influence on policies related to technology development and research.

3.1.2   The link between new technologies and the rise of new authoritarian regimes?

One can raise the question of why these new technologies are so important in today's geopolitics. Wright highlights three reasons why AI and the development of new technologies are critical to future development: (1) the fear of the so-called singularity stems from the assumption that there would be a point at which AI would exceed human capabilities; (2) the other worry is closely related to the first and draws our attention to the labor market, where these machines, robots, could easily replace humans; and (3) he also argues that fulfilling the second argument also has geopolitical consequences. By relying on AI, it is easier to monitor and control citizens. Wright puts it this way:

> …AI will offer authoritarian countries a plausible alternative to liberal democracy, the first since the end of the Cold War. That will spark renewed international competition between social systems.
>
> (Wright, 2019)

The first two aspects are closely related to economics; so at this point we will focus only on the third argument, which has a geopolitical significance. As we understand it, there are two crucial contradictions in this argument.

Wright argues that there is a choice between liberal democracy and a digital authoritarian state, but as we understand it, any kind of political regime is bound to refine and optimize its way of controlling society as technology advances. The more advanced and complex the technology that the state bureaucracy can use when monitoring the polity, the more comprehensive the surveillance can be conducted, and none of the political regimes can give away this advantage. In other words, the rise of AI and the proliferation of 5G, the Internet of Things, etc., also have profound implications for the relationship between state-citizen relationship in the West and China. Empirically, it is not a fait accompli that the proliferation of technology will necessarily produce authoritarian states, but it is clear that the relationship between citizens and the state is being reshaped by technology beyond our imagination, and the price of greater security and protection could be a significant loss of our privacy and freedoms. The 9/11 and the subsequent war on terrorism, the rise of ISIS, and the migration crisis in Europe have clearly demonstrated that new methods and approaches are needed in the West to control different versions of extremism, and that new technologies must be used not only in production but in every aspect of life, including the protection of citizens, to ensure security.

This line of argument presents the political contest over new technology as a heroic struggle between liberal democracies and authoritarian regimes, and fails to recognize that the trade war is also about a market competition fought by American companies to drive Chinese firms out of markets. Segal sheds light on this two-pronged interpretation:

> Huawei does pose a threat to US security, but that is not the only reason for Washington's assault on the company. Rather, the moves are a gambit in a larger battle over the future of the digital world.
>
> (Segal, 2019)

The perception that the debate over Chinese technology and the risks involved is about geoeconomic aspects as well as geopolitical ones can be confirmed by the increasing tensions between the EU and the United States, which have become more apparent with the Trump administration, too. Problems in transatlantic relations range from trade tariffs to the future of NATO, and obviously, the debate on how European powers treat China is another dividing line.

3.1.3 Fraying ties with Europe

There are both political and economic reasons for differences between the United States and the EU in dealing with China. The EU is weak as far as political and military power is concerned, but this same weakness and the resulting power asymmetry between the United States and the EU allows the United States to push the EU toward a stronger stance on change, while Europe prefers to focus on economic issues.

1   *Political reasons.* The asymmetry of power between the United States and the EU lies at the root of transatlantic tensions. This asymmetrical relationship was long accepted and cultivated by both sides after World War II (Poyakova & Haddad, 2019: 109–120); however, "the pivot to Asia," 9/11, and especially the rise of China have downgraded the role of Europe in American foreign policy, and the cultivated asymmetry between the United States and Europe does not seem acceptable to Washington, mainly due to the costs the American side has had to bear along the way. Recent American efforts to pressure European NATO members to adopt more financially responsible defense and security policies have provoked a backlash in European capitals. However, there is a catch in American efforts to share the financial burden: the more (financially) responsible European NATO countries bear the cost of their defense and security, the more sovereign foreign policies could be implemented by them; they might pursue a foreign policy that is not necessarily in line with American interests (Segal, 2019).

The Huawei case reveals this contradiction between the United States' two-pronged approach: bear the cost of your own security but follow my lead! The blow to US efforts to cut Huawei completely from the European market came from Germany and other European countries, which did not automatically exclude the Chinese company from the market. Moreover, the European Commission recommendation on cybersecurity (European Commission, 2019) falls short of a general ban of Huawei in the EU. However, there is a momentum clearly suggesting that European countries are taking an increasingly tough stance against Huawei: first, European allies seemed reluctant to follow US advice to ban Chinese 5G products and services, and then US foreign policy made significant efforts to technologically disconnect China from European countries and basically the rest of the world.

2   *Economic reasons.* In addition to foreign policy issues, economics also plays a role. Goldman points out some economic factors. He argues that the European competitors (Ericsson or Nokia) do not have the capacity in terms of research to replace the giant Huawei in this segment, and he adds that the products of Ericsson, Nokia, and Huawei are so intertwined that a complete ban of Huawei from the European market could not be done without affecting the other two European companies (Goldman, 2019). It should not be forgotten that American companies also have a significant influence on political decisions and vice versa. When the American President targets WeChat and TikTok, American business interests are also at stake. To be fair, if China bans certain American services (Google, YouTube, Facebook, etc.), we should not be too surprised if the United States does the same. Looking at the data (Table 3.4 shows the depth of the trade ties between China and the United States/the EU), we can see that the economic linkages with China are the same in both the United States and the EU. Thus, the

Table 3.4 Trade interdependency in EU-China and US-China relations in 2019

	Export	Import	Trade	Balance of goods and services (billion)
United States*	China is the third-largest market for the United States	China is the largest supplier of goods and services to the United States	China is the third-largest partner of the United States	$346
EU**	China is the third-largest market for the EU	China is the largest supplier of goods and services to the EU	China is the second-largest partner of the EU (after the United States)	€164

Source: * US data are from Office of the United States Trade Representative. ** The EU data originate from the Eurostat.

different responses cannot be explained by higher or lower economic linkages alone. As we saw earlier, the American trade deficit with China is decreasing due to the sanctions against China.

The most significant achievement of US-EU foreign policy cooperation was the launch of the so-called "New Atlantic Dialogue" on China. Secretary of State Michael Pompeo accepted High Representative Borell's proposal at German Marshall Fund's Brussels Forum in June 2020 to create a US-EU dialogue on China. In the speech, he listed the issues that should be addressed jointly:

- "People's Liberation Army's provocative military actions";
- "the CCP has broken multiple international commitments, including those to the WHO, the WTO, the United Nations, and the people of Hong Kong";
- "CCP's predatory economic practices, such as trying to force nations to do business with Huawei, an arm of the Chinese Communist Party's surveillance state";
- "violations of European sovereignty, including its browbeating of companies like HSBC";
- "Beijing's legion of human rights abuses"; and
- "the CCP's cover-up of the coronavirus" (Pompeo, June 25, 2020).

As can be seen, the list is long. The most important part of it, however, was when he addressed the Germans' greatest fear, the theft of intellectual property:

> Make no mistake, the Chinese have stolen a lot of German secrets, and the German people are worse off for that. Billions of dollars of intellectual property stolen by the Chinese Communist Party, outside of Germany. The hardworking German people created that intellectual property, worked hard for that intellectual property, built that, protected it in their system, and the Chinese came and stole it. And they've done it all across Europe and they continue to do it; they're doing it in the United States as well.
>
> (Pompeo, June 25, 2020)

There are two strands of foreign policy that we can see in this speech. The first element is the strong emphasis on intellectual property to engage the main European partner, Germany. The second element is the bellicose tone of the speech, reminiscent of the Cold War era. Fareed Zakaria reminds us of the mistakes that American foreign policy has made since 1945. According to him, this misguided policy led to the McCarthy era, nuclear armament, a long and pointless war in Vietnam, and several military interventions in the so-called Third World (Zakaria, 2020). The question that will be addressed in the next subchapter is how enticing these two strands of American foreign policy can be for Germany.

3.2 The core interests of German industry

The next subchapter focuses on an analysis of German economic interests in cooperation with China. This part attempts to interpret the business and economic development aspects of German industry in the light of German foreign policy. The main result is that Germany benefits more than the United States from China's growing economic presence in the world and in Europe because trade is not so imbalanced and German investment positions in China are also strong. At the same time, growing economic cooperation may also pose dangers for German companies, as Chinese acquisitions in Germany could threaten Germany's long-term economic and strategic interests.

3.2.1 German and Chinese trade ties

It is a triviality to say, but in this context it is more relevant than ever: Germany puts economic interests before politics in its bilateral relations with other countries. This statement also applies to relations with China. One of the reasons for this behavior is that China is extremely important to

Table 3.5 Trade interdependency in Germany-China and US-China relations in 2019

	Export	Import	Trade	Balance of goods and services (billion)
US	China is the third-largest market for the United States	China is the largest supplier of goods and services to the US	China is the third-largest partner of the US	$346
Germany	China is the fourth-largest market	China is the largest supplier of goods and services to Germany	China is the largest partner of Germany	€14

Source: US data are from the Office of the US Trade Representative, German data originate from the Federal Statistical Office (Statistisches Bundesamt).

Germany in both trade and investment relations. In 2019, China was the most important trading partner for Germany for the fourth year in a row. Trade volume was €206 billion, while trade with the United States was only €190 billion in 2019. Table 3.5 shows the trade linkages of Germany and the United States with China. The table illustrates the different trade and negotiating positions of Germany and the United States:

Germany also has a trade deficit with China, but unlike the American one, the size is not significant.

Moreover, China is "only" the fourth largest import market for Germany. In other words, Germany stands to gain more from a policy of engagement with China through deepening trade relations than the United States does in its relationship with China. The volume of trade is staggering, yet Germany is far from exhausting trade and business opportunities with China and can do so in a more balanced way than the United States.

3.2.2 Direct investments in Germany and China

Chinese investment increased significantly after China's accession to the World Trade Organization (WTO) in 2001 and again with the launch of the Going-Out Strategy in 2005, which drove the internationalization of Chinese firms. The last big push came after the Global Financial Crisis (2008–2009), and the peak of Chinese FDI outflows to Germany was reached in 2016. Since then, the appetite for Chinese investment waned for domestic political reasons, and this trend coincided with a new, stricter regulatory approach to Chinese investment in Germany.

Despite the surge in Chinese investment in Germany, the stock value of German investment in China is significantly higher than that of Chinese FDI in Germany. According to the data of German Federal Bank, the German FDI stock in 2019 was about €86 billion, accounting for 6.7 percent of the total German FDI abroad, and the Chinese FDI stock was about €5 billion (including Hong Kong),[1] so the Chinese share in 2018 was 0.9 percent of the foreign assets in Germany.

At the same time, we should add that the data—based on a balance of payments (BOP) approach—are not able to capture investment flows from offshore financial centers. Therefore, there are other datasets (e.g., the Mercator Institute for China Studies or the China Global Investment Tracker compiled by the American Enterprise Institute) that use a different approach to capture data on Chinese FDI in Europe and other regions. These combined annual transaction values tend to be much higher than the datasets using the approach of BOP, as they trace the investment back to the owner and do not take into account returns to China. In this analysis, we use the data retrieved using the BOP approach, as the alternative methods do not gather data in China and cannot be used for comparison.[2]

Comparison shows that the reciprocity in investment relations often demanded by the Germans is not achieved at the microeconomic level given the restrictions China imposes on FDI,[3] but at the macroeconomic level German FDI exceeds Chinese FDI in Germany, not only in the value of investment but also in the number of employees, firms, and annual turnover (see Table 3.6).

Data from American Enterprise Institute can be used to look at the structure of Chinese investment in Germany. From this data it can be concluded that Chinese firms invested mainly in the transport sector, that is, in the technology-intensive automotive firms. More than half of the $17 billion was concentrated in the 15 percent stake acquisition in Daimler ($11.8 billion). A similar concentration can be observed in the technology sector, where 87 percent of the funds spent in this sector were used to purchase KUKA, a company specializing in industrial robots. The acquisition of the Chinese company Guangdong Midea drew media attention to rising Chinese investment and was hotly debated in Germany. Mozur and Ewing summarize the story as follows:

> In Germany, the takeover of Kuka—frequently cited by politicians as emblematic of the country's future economic development—has drawn particular attention. The economics ministry (sic) examined the takeover of the company by Midea Group in China, which already owns 95 percent of Kuka shares, but eventually decided the deal did not meet the strict criteria for a formal review.
>
> (Mozur & Ewing, 2016)

Table 3.6 Chinese FDI in Germany and German FDI in China in 2018

Broken down by country of investment	Investments (€ billion)	Number of employees	Number of firms	Annual turnover (€ billion)	Main sector by country of investment and economic activity of domestic investors
Primary and secondary German FDI in China[a]	86	776,000	2,287	316	Automotive industry: 38 percent in terms of investment value
Primary and secondary German FDI in Hong Kong[a]	3.7	33,00	465	27	Financial industry: 41 percent in investment value
Chinese primary and secondary FDI in Germany	3.1	55,000	155	6.7	Manufacturing: 38 percent in term of investment value
Hong Kong primary and secondary FDI in Germany	2	10,000	108	5.5	–

Source: German Federal Bank Statistics (2020).
a Excluding Hong Kong.

The concentration of Chinese investment in two important economic sectors in Germany (automobiles and technology) is one of the main concerns of German politicians. German perceptions of China's role in foreign policy are complex, as they perceive China as an important partner in trade and China is assessed as an important destination for German direct investment, and yet Berlin is reluctant to recognize the role that Chinese firms could play for the German economy. At the same time, we must point out that the frequency with which the German Chancellor has visited China clearly shows that the German political elite is aware of China's economic relevance for German industry.

The temporal distribution of Chinese investment transactions in Germany shows no pattern which would make Chinese investment in Germany somehow special (see Table 3.7). According to the American Enterprise Institute, 68 larger FDI transactions were carried out by Chinese companies in Germany between 2007 and 2020. Similar to other European countries, the peak in the number of transactions was in the years 2015, 2016, and 2017, the three-year account for circa 68 percent of all Chinese direct investment between 2007 and 2020 (see more details in Table 3.8).

Table 3.7 Sectoral distribution of Chinese investments in Germany between 2005 and 2020[a]

Sectors	US$ (million)	Share (%)
Transport	21,820	45.86
Real Estate	6,460	13.58
Technology	5,910	12.42
Finance	3,710	7.80
Energy	3,640	7.65
Other	2,410	5.07
Health	1,400	2.94
Tourism	780	1.64
Metals	680	1.43
Logistics	440	0.92
Utilities	220	0.46
Entertainment	110	0.23
Total	47,580	100.00

Source: Own compilation based on American Enterprise Institute's dataset "The China Global Investment Tracker."
a   The data set was updated in early 2020.

Table 3.8 The temporal distribution of Chinese investments in Germany between 2007 and 2020[a]

Year	The number of transactions	The value of transactions (US$ million)	The share of transactions (%)
2007	1	130	0.27
2018	1	140	0.29
2009	0	0	0.00
2010	0	0	0.00
2011	5	2,040	4.29
2012	5	2,700	5.67
2013	3	800	1.68
2014	5	1,620	3.40
2015	6	1,690	3.55
2016	17	12,580	26.44
2017	11	7,560	15.89
2018	8	12,650	26.59
2019	5	5,530	11.62
2020	1	140	0.29
2007–2020	68	47,580	100.00

Source: Own compilation based on American Enterprise Institute's dataset "The China Global Investment Tracker."
a   The data set was updated in early 2020 (American Enterprise Institute, 2020).

3.2.3 Policy dilemmas

In the German social market economy (also called Rhine capitalism), the influence of business federations and trade unions on politics and economic development is more significant than it is in other advanced economies. That is why we should draw attention to the paper published by the Federation of German Industries in 2019, which uses the term "systemic competitor" for China. The phrase is very close to the term "systemic rival," used a few months later in a communication from EC (Federation of German Industries, 2019).

The position of the German government is unique for several reasons, not only with regard to relations with China. The country has more to lose from a deterioration of bilateral relations with China than other advanced countries in the EU (such as France, Italy, the Netherlands, and the Scandinavian countries). The challenges posed by China's economic development model could hit German companies harder than other EU member states, as the automotive industry is the backbone of manufacturing in both cases. Noah Barkin rightly points to Germany's hesitant attitude in shaping future relations with China:

> Germany's challenge in 2020 is to define a third space for itself and for Europe in the face of this growing US-China discord. But the Merkel government's reluctance to antagonize Beijing risks undermining the EU's push for a common policy toward China and perpetuating a situation where member states look out for their own interests, often to the detriment of a common European front.
>
> (Barkin, December 30, 2020)

We should not forget however that there is still no common or joint foreign policy in the EU. The so-called Common Security and Foreign Policy (CFSP), established in 1993, does not replace the individual foreign policies of the EU member states or Germany. A so-called common European front could only be formed if the interests were the same, and in this case the economic interests are different; the less developed countries of the EU could benefit more from Chinese direct investments and technology transfer, while the advanced countries of the EU could lose more, as they are the main targets of Chinese strategic investments and are threatened with technology theft.

Although calls for a tougher stance on China have grown louder in the EU, Berlin is not willing to risk bilateral relations with Beijing just to present a united front.[4] But the European push against China does not necessarily clash with the goals of German foreign policy. The hesitation of other European countries and Germany's willingness to compromise with China can be a powerful mix to achieve German goals. Barkin cites a German officer in this context:

> Merkel has no problem with pushback against China as long as it's not Germany that is doing the pushing.
>
> (Barkin, December 30, 2020)

Angela Merkel has made 12 trips to China in her 15 years as German Chancellor, a testament to engagement, but to grasp the complexity of China-Germany relations, it is also important to note the growing tensions between

Germany and the United States. During the Trump administration, the fierce debate over sharing the financial burden of NATO severely strained German-American relations. The strain on these relations caused by the "America First" approach—which stands in sharp contrast to the logic that the hegemonic power must take responsibility for the provision of global public goods—may easily change with a new administration starting in 2021 and reduce the looming bilateral problems between the United States and Germany. However, the fundamental shifts that are redefining the role that the EU and Germany can play in global politics and the global economy are still in place, and the simmering tensions will not subside until Germany can find a new role for itself in the EU and in the world. Although there is growing pressure on Germany to take a tougher stance on China (Barkin, March 8, 2020), questions remain as to whether Germany can stand up for its fundamental interests—as we saw above—to deepen its trade and investment ties with China.

3.3 The balancing Russia

Among the three countries influencing China's growing economic presence in Europe, Russia is the weakest economic power. While it would make sense for Russia to cooperate with China on many levels, China seems interested only in exporting Russian mineral resources and military technology. The real unifying element between the two countries is the fact that the rivals are the same in both politics and economics. The logic of "the enemy of my enemy is my friend" can only be temporary.

3.3.1 From "Greater Europe" to "Greater Eurasia"

Before diving into the details of technological cooperation with China, it must be made clear that Russia's position is completely different from Germany's, since Russia is not part of the transatlantic institutional framework (EU, NATO) and the country is really a Eurasian complex because of its history. Being economically backward, it also needs technology transfer from other countries. In many ways its economic situation is reminiscent to that of the countries of Central European, as it also needs capital and technology imports; however, Russia is a leader in military technology and because of its size can accumulate capital and create a huge domestic market.

After the country's failed (or rather, not even really attempted) integration into the Western framework and the annexation of Crimea, the concept of "Greater Asia" was replaced in Russian foreign policy thinking in 2014 by the concept of the "Greater Eurasian Partnership." Several analysts point to the influence of Alexander Dugin on Russian political thinking, whose ideas question the morality of a unipolar world order, noting:

> When there is only one power which decides who is right and who is wrong and who should be punished and who not, we have a form of

global dictatorship. This is not acceptable. Therefore, we should fight against it. If someone deprives us of our freedom, we have to react. And we will react. The American Empire should be destroyed. And at one point, it will be...

(Dugin, 2012: 193)

Despite the mass media coverage of Dugin, we cannot take Dugin's direct influence on Russian foreign policy for granted. He argues that China poses a threat to Russia, which should seek support from South Korea, Japan, Vietnam, and India to counterbalance this rising power (Radin & Reach, 2017), while the "Russian pivot" to Asia can be framed as a pivot to China. This shift in foreign policy was the result of long-term shifts in power toward Asia (i.e., China); however, the shift was accelerated by Western sanctions against the country following the annexation of Crimea. As in the case of the United States, the swing toward Asia began long before it became obvious to observers: Russia joined the Asia-Pacific Economic Cooperation (APEC) in 1998; it cofounded Shanghai Cooperation Organization along with China, Kazakhstan, Kyrgyzstan, Russia, Tajikistan, and Uzbekistan; and in 2015 Russia established the Eurasian Economic Union (EAEU), which includes five countries (Russia, Belarus, Armenia, Kazakhstan, and Kyrgyzstan). As a result of this long process, Russia became a power that, on the one hand, belongs to the periphery of the world economy and, on the other hand, is still a significant power in world politics due to its military capabilities and technology, a country that can forge alliances and strategic partnerships on its own.

3.3.2 Potential in technology cooperation with China

Russian interests in technology cooperation with China stand in stark contrast to those of Germany. There are two main reasons for the significant differences:

1   The benefits to be gained from direct Chinese technology transfer are hypothetically greater, and the very logic of "the enemy of my enemy is my friend" tells us that Chinese and Russian technology cooperation must be intense. At the same time, as we will see later, China has not made significant investments in technology-related areas, and Russia is not interested in handing over its military technology to the Chinese.

2   The sanctions against Russia and the trade war against China should make the two Eurasian powers natural partners on many issues and could even make them allies, but the long history of deep mistrust blocks further steps despite official pronouncements by politicians to the contrary. Moreover, China's growing economic and political influence in Central Asia, which Moscow considers its "near abroad" or a Russian sphere of influence, keeps Russia alert.

These problems still overshadow bilateral relations between these countries, although political and economic cooperation intensified between 2010 and 2014, and this trend has accelerated since 2014, when the United States and the EU imposed restrictions on Russia after the Crimean Crisis. Despite the relatively good relations between the two countries, China was not willing to offer financial assistance to Russia when the economic impact of the sanctions led to an economic crisis in the country (Gorenburg, 2020).

Regardless of this backlash, there is significant potential for technological cooperation between the two countries. In late 2019, the two countries developed a roadmap for the so-called "Year of Russian-Chinese Scientific, Technical and Innovation Cooperation," the intention of which was reaffirmed by Russia signing the decree by the Russian President, making 2020 the "Year of Russian-Chinese Scientific, Technical and Innovation Cooperation." The focus of cooperation is set on AI, communications, and the Internet of Things (Elmer, 2019). Scientific and technological cooperation is closely related to military cooperation between the two countries. Especially in the military field, this cooperation is not free of friction. Elmer puts it this way:

> However, some Russian officials are frustrated that China is catching up its technology and have accused its neighbour of infringing its intellectual property. Yevgeny Livadny, chief of intellectual property projects of Russian defence conglomerate, said earlier this month that there was a 'huge problem' with China copying Russian aircraft engines, planes and other defence system, which is a 'huge problem'.
>
> (Elmer, 2019)

Russian and Chinese technological cooperation takes various forms, ranging from scientific institutes to private firms and even including cooperation in the field of military technology. However, Samuel Bendett and Elsa Kania indicate that this cooperation has recently shifted toward the creation of joint platforms, demonstrating the need to find more flexible forms of cooperation through which public actors can still direct private interests by setting specific goals for research cooperation. A few examples of such collaborations:

1   The Russian Foundation for Basic Research and National Natural Science Foundation of China launched a joint competition in mathematics, physics, chemistry, and biology.
2   Huawei opened a research lab in Moscow and plans to open a Huawei Academy of Information and Communication Technologies and train Russian developers.
3   At the Sino-Russian Engineering Technology Forum, 15 projects were signed in 2019, valued at nearly $1.1 billion (Bendett & Kania, 2020).

3.3.3 Unbalanced trade and investment relations

As noted earlier, tensions fester in Sino-Russian relations. The tensions are fueled by severe imbalances in trade and investment, leading to incidents between the two countries. One of the most recent incidents was that Valery Mitko, the president of the St. Petersburg Arctic Social Sciences Academy, was charged with treason in June 2020. To observers, the details of this story may seem familiar, as similar charges have been a strong feature of American-Russian or British-Russian relations. The surprising element was that Mitko allegedly handed over classified material to Chinese intelligence (TASS, 2020). This report warns us not to exaggerate the potential for Chinese-Russian technological cooperation. Russia is aware of the imbalances and long-term trends that place China in an increasingly dominant position. To demonstrate the unbalanced position in research and development (R&D), it is enough to look at China's and Russia's spending on R&D: in 2019, China spent almost ten times ($533 billion) more on R&D than Russia ($61 billion).[5] Russia is willing to cooperate with China, but some relationships work better than others. Korolev argues:

> … since the end of the Cold War, China and Russia have constructed comprehensive strategic cooperation, of which some aspects are more advanced than others. However, all aspects have consistently progressed over the last two decades.
>
> (Korolev, 2019: 24)

He points out that military cooperation is the most advanced aspect of such cooperation, involving joint military exercises as well, while in diplomacy, they do not always behave as allies, although elements of support in international organizations can easily be found. Economic cooperation is the weakest element, and the main emphasis is still put on energy cooperation. The two countries are on an unequal basis in trade, which can easily be shown.

The EU was still Russia's largest partner in 2019, with $260 billion in trade, but trends and the geopolitical constellation—EU economic sanctions against Russia are still in place—favor China, whose trade totaled $110 billion, surpassing Germany for the first time in 2019. In 2018, trade with Russia accounted for 0.8 percent of China's total trade volume, while China's share of Russian trade was a whopping 15.5 percent (Hillman, 2020: 2).

Based on this data, it can be argued that opening the economy to Chinese goods and services would be a self-defeating strategy, and the Russian government seems to be aware of this trap. This realization is reflected in the EAEU-China trade agreement, which aims not to reduce tariffs but to facilitate trade by reducing bureaucratic costs and simplifying trade rules. At the same time, trade protectionism cannot be the way forward for Russia either, as inward-looking economic policies have never delivered sustainable growth and could only provide a temporary respite from the social costs of adjustment and modernization.

Table 3.9 Sectoral distribution of Chinese investments in Russia[a]

Sectors	US$ million	Share (%)
Energy	30,850	55.70
Metals	65,00	11.73
Real Estate	4,850	8.76
Chemicals	3,490	6.30
Transport	2,650	4.78
Agriculture	2,480	4.48
Other	2,340	4.22
Finance	1,200	2.17
Technology	880	1.59
Logistics	150	0.27
Total	55,980	100.00

Source: Own compilation based on American Enterprise Institute's dataset "The China Global Investment Tracker."
a   The data set was updated in early 2020 (American Enterprise Institute, 2020).

Chinese investment is desperately needed in Russia, whose infrastructure could be significantly improved by Chinese capital investment, but it is difficult to systematically assess Chinese Belt and Road Initiative investment at this point since the number of project announcements does not correlate with the number of de facto projects implemented. Even in this case, we have to rely on China Global Investment Tracker, which also publishes sectoral investment data. Between 2015 and 2020, the total amount of Chinese investment was $54 billion; this amount is only slightly higher than the Chinese investment in Germany ($47.6 billion) and significantly lower than the investment in the United Kingdom ($86.4 billion). Another significant difference in this case is the very clear emphasis on the energy sector and an insignificant share of Chinese investment in technology-related sectors. The difference becomes even more significant when comparing Russia's share of technology-related investment of 1.59 percent with Germany's figure of 12.42 percent (Table 3.9 shows the sectoral distribution of Chinese investments in Russia).

3.3.4 Policy dilemmas

We can see that the relationship between Russia and China is extremely unbalanced and is becoming more unbalanced as time goes on. Not only is Russia trying to balance American and European sanctions by relying more on China, but it is also attempting to diminish China's economic and political role by building relations with India. See, for example, Russian arms sales to India in the summer of 2020, even after the border disputes between China and India (Siow, 2020). We could conclude that Russia would achieve this balance by strengthening economic ties with the EU; however, at this point, being isolated from the West, Moscow has no options other than deepening ties with Beijing.

The stakes are high when it comes to technology transfer, as Russia remains a frontrunner in military technology, while China's achievements in manufacturing, 5G, and AI R&D make it one of the most competitive nations in the world. Unlike Germany, Russia has long-term and unresolved geopolitical conflicts of interest with China:

- the struggle for influence in Central Asia;
- the fear of Sinification of the Russian Far East, which resurfaces from time to time; and
- the fear that China might gain a significant market share of global arms sales.

Maria Siow cites Alexey Muraviev, an associate professor of Australia's Curtin University:

> The Chinese are also engaged in reverse engineering Russia's military technology and then trying to sell indigenous platforms based on Russian designs, thereby competing against Russia on the global arms sales market...
>
> (Siow, 2020)

Despite the supposedly strong ties between China and Russia, China is only a second bet for Russia, and in the case of Russia's improving relations with the EU and the United States, there is no question which one Russia would choose while retaining its "sovereign niche" in world politics. The expected results of Chinese cooperation make it tempting for Russia to deepen cooperation; however, China's romance with the Central European countries also shows that failure to deliver on promises (insignificant amounts of investment, only a few attempts to channel new technology to these countries) can easily end a romantic relationship. Emilian Kavalski puts it this way:

> Most CEE states have been socially distancing from Beijing even before the pandemic. The adage that it takes two to tango, applies to international politics as well. Rather than a long-term partner, China appears to have been a fleeting interest whose appeal quickly wore off.
>
> (Kavalski, July 1, 2020)

The irony of this situation is that even China has stronger reasons to cooperate economically with the EU than with Russia. However, China has never had any real geopolitical interests in Europe, while the Eurasian continent with its vast energy and natural resources is important for China's catch-up process.

In this chapter we have seen what the main economic interests of the United States, Russia, and Germany are when it comes to cooperation with China. We were able to understand why the United States tends to view China as a rival rather than Germany, which has relatively the most to gain from this relationship. And thus, we can also explain Germany's more supportive attitude toward China, while the warm relations between China and Russia could come to an abrupt end if the current status quo in EU-Russia or United States-Russia relations comes to an end. The next chapter focuses on the legal, business, political, and even technological environment that Chinese firms face in the three main EU economies: Germany, France, and Italy. This chapter focuses on Chinese direct investment in these countries; the sectoral distribution also shows business opportunities and latent geopolitical threats. This part also gives a brief overview of each country's endowment to participate in R&D, which helps us understand the room for maneuver in the technological race between the United States and China.

Notes

1 Sources like Statista and China National Official Statistics publish higher numbers for Chinese FDI stock in Germany (Statista: $12.1 billion in 2017; China National Official Statistics: $13.7 billion in 2018); however, we use this comparison since the German Federal Bank publishes German FDI in China too in the same breakdown (country of origin).
2 According to the Mercator Institute, the combined value of Chinese FDI in Germany was €22.7 billion between 2000 and 2019 (Kratz et al., 2020), while the American Enterprise Institute estimates that Chinese firms invested $47.6 billion in Germany between 2005 and 2020 (American Enterprise Institute, 2020).
3 European firms can enter the Chinese market only when establishing joint ventures with Chinese firms, and in some sectors FDI is entirely blocked. The Foreign Investment Law came into force on January 1, 2020, and it brought several significant changes; however, they only apply to the newly established firms. At the same, the negative list containing the blocked sectors for foreign investors was updated in mid-2020. According to Zhou:

> Compared to the 2019 negative lists, the new 2020 National Negative List has cut the number of restrictive measures by 17.5 percent from 40 to 33, and the new 2020 FTZ Negative List has cut the measures by 18.9 percent from 37 to 30.
>
> (Zhou, 2020)

4 In this aspect, the French approach is different. Macron called for a different approach in the case of China ahead of the China-EU summit in 2019: "Since the beginning of my mandate, I have been calling for a real awareness and defence of European sovereignty … Finally, on subjects as important as China, we have it, …" (AFP, 2019).
5 The data are measured based on purchasing power parity and originate from the R&D World Magazine.

References

AFP (2019, March 21). Macron Hails Europe 'Awakening' to China Threat. *AFP*, Retrieved from: https://www.france24.com/en/20190321-macron-hails-europe-awakening-china-threat

American Enterprise Institute (2020). China Global Investment Tracker. Updated in 2020, *AEI*, Retrieved from: https://www.aei.org/china-global-investment-tracker/

Barkin, N. (2020, December 31). What Merkel Really Thinks About China—and the World. *Foreign Policy*, Retrieved from: https://foreignpolicy.com/2020/12/31/what-merkel-really-thinks-about-china-and-the-world/

Barkin, N. (2020, March 8). Why Post-Merkel Germany Will Change Its Tune on China. Pressure Is Building in Berlin to Get Tough on Beijing. *Politico*, Retrieved from: https://www.politico.eu/article/why-post-merkel-germany-will-change-its-tune-on-china/

Bendett, S. & Kania, E. (2020, August 12). The Resilience of Sino-Russian High-Tech Cooperation. War on the Rocks. *Texas National Security Review*, Retrieved from:https://warontherocks.com/2020/08/the-resilience-of-sino-russian-high-tech-cooperation/

Clinton, H. (2011, October 11). America's Pacific Century. *Foreign Policy*, Retrieved from: https://foreignpolicy.com/2011/10/11/americas-pacific-century/

Donnan, S. & Hamlin, K. (2018, November 11). These Products Show How Hard It'll Be to Beat China in Trade War. *Bloomberg*, Retrieved from: https://www.bloomberg.com/news/articles/2018-11-11/trump-s-china-cold-war-yields-hard-look-at-global-supply-chains

Dugin, A. (2012). *The Fourth Political Theory*. London: Arktos

Economy, E. C. (2020, March 13). Exporting the China Model. Prepared Statement by Elizabeth C. Economy. In Council on Foreign Relations. *Testimony before the U.S.-China Economic and Security Review Commission Hearing on The "China Model"*, pp. 24–32. Retrieved from: https://www.uscc.gov/sites/default/files/2020-10/March_13_Hearing_and_April_27_Roundtable_Transcript.pdf

Edelstein, D. M. & Krebs, R. R. (2015, November/December). Delusions of Grand Strategy. The Problem with Washington's Planning Obsession. *Foreign Affairs*, pp. 109–116, Retrieved from: https://www.foreignaffairs.com/articles/2015-10-20/delusions-grand-strategy

Elmer, K. (2019, December 28). China and Russia Plan to Boost Scientific Cooperation with Focus on Artificial Intelligence and Other Strategic Areas. *South China Morning Post*, Retrieved from: https://www.scmp.com/news/china/diplomacy/article/3043787/china-and-russia-plan-boost-scientific-cooperation-focus

European Commission (2019). Commission Recommendation of 26.3.2019. Cybersecurity of 5G Networks. *Strasbourg*, Retrieved from: https://ec.europa.eu/digital-single-market/en/news/cybersecurity-5g-networks

Federation of German Industries (2019, January). Partner and Systemic Competitor—How Do We Deal with China's State-Controlled Economy? *BDI*, Policy Paper, Retrieved from: https://www.wita.org/wp-content/uploads/2019/01/201901_Policy_Paper_BDI_China.pdf

Goldman, A. P. (2019, May 20). US ban won't derail Huawei's European 5G rollout. *Asia Times*, Retrieved from: https://asiatimes.com/2019/05/us-ban-wont-derail-huaweis-european-5g-rollout-2/

Gorenburg, D. (2020, April). An Emerging Strategic Partnership: Trends in Russia-China Military Cooperation. *George C. Marshall*, European Center for Security Studies, No. 054, Retrieved from: https://www.marshallcenter.org/en/publications/security-insights/emerging-strategic-partnership-trends-russia-china-military-cooperation-0

Hillman, J. E. (2020, July 15). China and Russia: Economic Unequals. *Center for Strategic and International Studies*, Retrieved from: https://www.csis.org/analysis/china-and-russia-economic-unequals

Kavalski, E. (2020, July 1). The End of China's Romance with Eastern Europe? *Global Policy*, Retrieved from: https://www.globalpolicyjournal.com/blog/01/07/2020/end-chinas-romance-eastern-europe

Korolev, A. (2019, May). How Closely Aligned Are China and Russia? Measuring Strategic Cooperation in IR. *International Politics, 57*, pp. 760–789.

Kratz, A., Huotari, M., Hanemann, T. & Arcesat, R. (2020, April 08). Chinese FDI in Europe: 2019 Update. *Rhodium Group (RHG) and MERICS*, Retrieved from: https://mimderics.org/en/report/chinese-fdi-europe-2019-update

Mehta, A, (2017, November 2). Google's Schmidt: US Losing Edge in AI to China. *C4ISRNET*, Retrieved from: https://www.c4isrnet.com/it-networks/2017/11/02/china-on-path-to-eclipse-us-with-ai-warns-google-head/

Mozur, P. & Ewing, J. (2016, September 16). Rush of Chinese Investment in Europe's High-Tech Firms Is Raising Eyebrows. *New York Times*, Retrieved from: https://www.nytimes.com/2016/09/17/business/dealbook/china-germany-takeover-merger-technology.html

Pamuk, H. & Shalal, A. (2020, May 4). Trump Administration Pushing to Rip Global Supply Chains from China: Officials. *Reuters*, Retrieved from: https://www.reuters.com/article/us-health-coronavirus-usa-china-idUSKBN22G0BZ

Pompeo, M. R. (2020, June 25). A New Transatlantic Dialogue. Speech. Washington. *German Marshall Fund's Brussels Forum*, Retrieved from: https://useu.usmission.gov/secretary-pompeos-remarks-at-brussels-forum/

Poyakova, A. & Haddad, B. (2019, July/August). Europe Alone. What Comes After the Transatlantic Alliance. *Foreign Affairs*, pp. 109–120, Retrieved from: https://www.foreignaffairs.com/articles/europe/2019-06-11/europe-alone

Radin, A. & Reach, C. (2017). *Russian Views of the International Order. Building a Sustainable International Order.* A RAND Project to Explore U.S. Strategy in a Changing World, Santa Monica, CA: RAND Corporation.

Richter, F. (2020, February 25). These Are the Top 10 Manufacturing Countries in the World. *World Economy Forum*, Retrieved from: https://www.weforum.org/agenda/2020/02/countries-manufacturing-trade-exports-economics/

Rotondi, R., Skolimowski, P., Neumann, J. & Lima, J. (2020, June 30). Europe Finds It's Not So Easy to Say Goodbye to Low-Cost China. *Bloomberg*, Retrieved from: https://www.bloomberg.com/news/articles/2020-06-29/europe-finds-it-s-not-so-easy-to-say-goodbye-to-low-cost-china

Segal, A. (2018, September/October). When China Rules the Web. Technology in Service of the State. *Foreign Affairs*, Retrieved from: https://www.foreignaffairs.com/articles/china/2018-08-13/when-china-rules-web

Segal, A. (2019, July 11). The Right Way to Deal with Huawei. The United States Needs to Compete with Chinese Firms, Not Just Ban Them. *Foreign Affairs*, Retrieved from: https://www.foreignaffairs.com/articles/china/2019-07-11/right-way-deal-huawei

Siow, M. (2020, August 22). Could Russia Side with the US and India Against China? *South China Morning Post*, Retrieved from: https://www.scmp.com/week-asia/politics/article/3098398/could-russia-side-us-and-india-against-china

TASS (2020, June 15). Russian Arctic Academy President Accused of Working for Chinese Intelligence. *TASS. Russian News Agency*, Retrieved from: https://tass.com/society/1167581

Van den Bossche, P., et al. (2020). Trade War Spurs Sharp Reversal in 2019 Reshoring Index, Foreshadowing Covid-19 Test of Supply Chain Resilience. *Kearney*, Retrieved from: https://www.kearney.com/operations-performance-transformation/us-reshoring-index

Vanden Heuvel, K. & Cohen, S. F. (2015, October 28). Edward Snowden: A 'Nation' Interview. *The Nation*, Retrieved from: https://www.thenation.com/article/archive/snowden-exile-exclusive-interview/

Wright, N. (2019, July 10). How Artificial Intelligence Will Reshape the Global Order. The Coming Competition between Digital Authoritarianism and Liberal Democracy. *Foreign Affairs*, Retrieved from: https://www.foreignaffairs.com/articles/world/2018-07-10/how-artificial-intelligence-will-reshape-global-order

Zakaria, F. (2020, January/February). The New China Scare. Why America Shouldn't Panic About Its Latest Challenger. *Foreign Affairs*, Retrieved from: https://www.foreignaffairs.com/articles/china/2019-12-06/new-china-scare

Zhou, Q. (2020, July 1). China's 2020 New Negative Lists Signal Further Opening-Up. *China-Briefing*, Retrieved from: https://www.china-briefing.com/news/chinas-2020-new-negative-lists-signals-further-opening-up/

4 Chinese investment and 5G cooperation in the EU

France, Germany, and Italy

4.1 Germany: a case study

4.1.1 Introduction

The next part of the chapter focuses on the long-term motivations of the major European countries (Germany, France, and Italy) in relation to 5G issues and then looks at the sectoral distribution of Chinese investment in the four European countries selected. However, before going into the details of each country, it should be noted that the selected countries are at different stages of 5G readiness, so their approach to FDI in general and to Chinese investment differs. Looking at the three major economies of the EU, there are significant differences in their ability to benefit from 5G and economic cooperation with China. Table 4.1 shows the 5G Readiness Index of the three countries, prepared by a Luxembourg-based consulting firm (inCITES). According to the index, which includes six main categories (infrastructure and technology, regulation and policy, innovation landscape, human capital, country profile, and demand), Germany performed poorly in the demand category but was above the median in other categories. France scored above average in all categories, but not significantly so,[1] so its overall index was slightly lower than Germany's. At the same time, Italy performed the worst, with the country's scores below average in the areas of regulation and innovation.

5G investment is seen as a strategic area all over the world, so regulation of this area is essential to strengthening national sovereignty. The regulation

Table 4.1 5G Readiness Index[a] of Germany, France, and Italy

Germany	66.68	3
France	59.38	13
Italy	53.01	20

Source: InCITES.

a There are six main categories selected for the 5G readiness index by inCITES Consulting: infrastructure and technology; regulation and policy; innovation landscape; human capital; and country profile demand.

DOI: 10.4324/9781003128625-4

of FDI in general and in this field as well as the sectoral distribution of Chinese investment in Germany help us to understand what is happening on the ground and what is the real story behind the newspaper headlines. For this reason, this subsection first examines the legal framework for foreign direct investment (FDI) from third countries (FDI screening), and then discusses the areas in which Germany underperforms or outperforms in technology-intensive sectors to see where the potential for cooperation between China and Germany can be observed.

4.1.2 The legal framework

The German legal framework for FDI screening is relatively liberal, but the attitude toward FDI from third countries has become tougher in recent years. FDI screening in Germany is governed by the Foreign Trade and Payments Act and its accompanying ordinance, the Foreign Trade and Payments Ordinance. The last amendment to the law took place in April 2020 and came into force in October 2020, and the ordinance was amended in the same year to include the health sector in the list of no-defense sectors that are specially protected (Engelstaedter & Gernoth, 2014).

Based on the new legal framework, the Ministry of Economic Affairs and Energy can review and prohibit an investment if the buyer is not a resident of the EU. The Ministry can investigate the investment if the acquisition of voting rights in the company is at least 25 percent in any sector. It is very important to emphasize that not only the direct investment but also the indirect participation of at least 25 percent of the voting rights can be screened and prohibited by the Ministry. In addition, the same law can be applied if the foreign buyer already owns a German resident company with a shareholding of at least 25 percent and this company acquires a third company in Germany. While the basic rule is that the threshold is 25 percent, in defense sectors and specially protected sectors the threshold has been lowered to 10 percent of the voting rights or assets (Federal Ministry for Economic Affairs and Energy, 2018). In the former case, the Ministry of Economic Affairs and Energy must be notified by non-German buyers, while in the latter case it must also be informed by EU foreign investors. Although 5G itself is not explicitly mentioned in the description of the defense and specially protected sectors,[2] the list of protected sectors allows for a flexible interpretation and the term "operators of critical infrastructure in telecommunication" can be used to investigate companies operating in the 5G sector.

The review procedure must begin within a period of three months after the transaction has been concluded. After receiving the necessary information and documents from the foreign buyer, the Ministry has a maximum of two months to complete the review procedure. On the one hand, the buyer is not obligated to inform the Ministry of the transaction, and on the other, the buyer may request a clearance certificate from the Ministry stating that the transaction does not pose a threat to public order or security. After receiving such a certificate or after the expiration of the two-month investigation period, the Ministry cannot prohibit the transaction.

4.1.3 Technology-intensive sectors

According to the Mercator Institute for China Studies (Kratz et al., 2020), Chinese companies invested about €22.7 billion in Germany between 2000 and 2019, while China Global Investment Tracker reports FDI stock of $47.5 billion (American Enterprise Institute, 2020). Regardless of the datasets we use, Germany is the EU member state where Chinese companies invested the most (until the United Kingdom left the EU in January 2020, it was the EU member state most affected by Chinese investment[3]). Despite the significant differences in overall volumes, both data collections show a strong downward trend in Chinese investment in Germany after peaking in 2016–2017.

The cumulative value of Chinese investment is not significant for the German economy, but its sectoral distribution could give rise to some concerns for German policy and decision-makers. Between 2005 and 2020, the technology sector accounted for about 13 percent and the transport sector almost 47 percent of Chinese investments. When it comes to the transport sector, these transactions mainly included investments in the automotive sector. Of the 25 Chinese investments in the transport sector, 22 were in the automotive sector and the rest in the aviation industry. A heavy concentration of Chinese assets can be observed in the automotive sector, with 47 percent of all Chinese investment in the automotive sector consisting of two transactions for the acquisition of 15 percent of Daimler. An even greater concentration is observed in the technology sector, where 87 percent of all Chinese investment in the sector consists of four transactions that secured 94.6 percent of the assets for Chinese Guangdong Midea.

The concentration of Chinese investment in two key sectors of the German economy (automotive and technology) is the focus of German policy. To this debate we can add the following aspects:

1 Germany's performance is mixed in the field of innovation. According to the Bloomberg Innovation Index 2020, Germany ranks first, breaking the dominance of South Korea. However, in the Swiss-based International Institute for Management Development World Digital Competitiveness Ranking 2019, Germany ranks 14th. This composite index is made up of three dimensions: knowledge (12th), technology (31st), and future-readiness (16th). Among the indicators, the weakness of digital skills, the number of broadband subscribers, and investment in telecommunications are highlighted (IMD, 2019: 78–79).

2 In this way, the Digital Economy and Society Index (DESI), published by EU, summarizes Germany's relative position in new technologies and future capability:

Germany ranks 12th out of 28 EU Member States in the 2020 edition of the Digital Economy and Society Index (DESI). Based on data prior to the pandemic, Germany performs well in most DESI dimensions, except in digital public services, where it ranks 21st. On the Connectivity dimension, Germany leads the EU on 5G readiness and has a high take-up of overall fixed broadband. However, performance in fixed very high capacity network coverage is below the EU average, where it ranks 21st.

(European Commission, 2020a: 3)

3   There are also areas, such as artificial intelligence (AI), where Germany is in danger of being left behind. Although Germany ranked fifth in Tortoise Intelligence's Global AI Index in 2019, there is a feeling among German policymakers that Germany may be a laggard in the AI race. This realization came when the German Chancellor visited the Chinese technology center in Shenzhen in 2018. She expressed her concerns as follows:

"For centuries, or let's say since the age of Enlightenment, we in Europe were used to being the first ones to come up with technological innovations," ... "That's not the case anymore today. And this should worry us.

(German Chancellor Angela Merkel,
cited by Delcker (July 23, 2018))

4   Germany is better at traditional technologies. The largest companies in Germany are simply much older than their Chinese or American counterparts. According to the Factbook of Digitization,[4] large German companies are on average 114 years old, while American companies are on average 30 years old and the largest Chinese companies are 34 years old. This "age difference" immediately shows that their specialization must be in different industries, where the technologies used must also be different (ESCP, 2019: 3).

5   The difference lies not only in the "age" of these companies, but also in their size, which is significantly smaller than that of their American counterparts. The market value of Amazon, Google, and Apple was greater than the market value of all (763) listed German companies combined in August 2019.

6   Technology companies with a market capitalization of more than $100 billion accounted for 8 percent of the market capitalization of the 50 most valuable companies in 2018, compared with 28 percent in the United States and 22 percent in China (Bird et al., 2018).

7   The "Factbook Digitalization" report points out the weakness of venture capital funding, which implies a small number of small high-tech firms

with great future potential. Between 1995 and 2018, the share of companies financed by venture capital amounted to 1.3 percent of GDP in Germany, compared to 34 percent in the United States (ESCP, 2019: 5).

8   Germany's strongest sector is the automotive industry, and in addition to traditional competitors, German companies must contend with newcomers in the production of electric cars (Tesla has the largest market value in this segment),[5] and the proliferation of car-sharing agencies could also limit the growth of these companies. In both cases, new technology is the key element of market change, and German automakers do not appear to be at the forefront of these new technologies.

Germany lags behind its main competitors in some soft but important indicators (see Table 4.2), but even Alibaba invested ten times more in AI in 2018 than the investments of German players/actors combined. As we have already seen, the strengths of the Chinese economy lie in manufacturing, 5G, and AI, so we can easily understand why several representatives of German industry have shifted the focus of policy and public debate to the alleged threats.

To sum it up, there is no German company that can compete technologically with the world leaders in 5G technology, be they Chinese, American, Korean, or other European companies. If the interventionist approach at the EU level to regulating the market gains the upper hand, Germany will have no choice but to fully support Ericsson and Nokia. However, if the market-friendly approach prevails in the EU—Germany certainly has the ability to influence the outcome of this debate—Germany could benefit from Chinese investment in this area. It seems that the two countries are largely in agreement on "realpolitik" issues, but on trade and investment relations there are differences of opinion that have yet to be resolved.

Ironically, what is happening to German industry today (new foreign capital, technology infusion, and companies entering the German market) is very similar to what happened three decades ago in Central Europe, when

Table 4.2 Selected indicators of German tech competitiveness

	Germany	*United States*	*Japan*
Share of fiber optics of all broadband connections	1.6%	14.3%	74.0%
Number of most powerful supercomputers	14	119	219
Investments in quantum computers in billion $	0.74	1.2	10

Source: ESCP (2019: 9).

German companies, who were the foreign buyers at the time, entered the market, bought up companies, and transformed the economies of Central Europe. Central European countries benefited from this process by importing technology, knowledge, and capital, and in some cases the newly established firms also brought with them strong international market positions for their subsidiaries, while the one-sided dependence of these economies on Germany has become stronger than ever. At the same time, this analogy is somewhat flawed, as the German economy and society are in much better shape than the countries of Central Europe at the time, and Germany can also benefit much more from the growing economic cooperation with China without falling into a one-sided relationship.

We argue that Germany has the most to lose and the most to gain by improving its relations with China. The dilemma is reflected in German perceptions of China's role in German foreign and trade policy, which are complex and open to "realpolitik" arguments. Germany perceives China as an important trading partner. The competitive nature of cooperation in manufacturing could be mitigated by adopting the EU-China Comprehensive Investment Agreement, and the complementary nature in 5G and AI could be strengthened by more cooperation between these two partners.

4.2 France: a case study

4.2.1 Introduction

France is the only country among EU members that has a geopolitical conflict of interest with China. China's rising economic power in Africa, especially in the former French colonies, is a source of concern for French politicians. Nonetheless, these conflicts can be easily resolved if other economic tensions can be defused. In our view, the real conflict lies in imbalanced trade and much less in investment relations.

1   The economic relationship between France and China is unbalanced in terms of trade and investment. Like many countries in the world, France has a significant trade deficit with China. Moreover, China is the largest source of the trade deficit for France (€34.4 billion in 2018), with about one-third of the trade deficit coming from trade with China (Table 4.3 shows France's trade balance with China and its relative importance in the overall balance).

2   At the same time, the French FDI stock is significantly higher than the Chinese FDI stock in France. In 2018, France's FDI stock in China amounted to €23 billion, while the Asian country invested only €8.6 billion in France. This €8.6 billion represented 1.2 percent of France's FDI stock abroad![6]

3   If we use data from the Mercator Institute, we have much higher figures for Chinese FDI in France.[7] According to data on the total value of

Table 4.3 French trade balance with China ($ billion)

	2010	2018	The overall balance in 2010	The overall balance in 2018
France	−34.3	−34.4	−87.5	−90.8

Source: World Bank WITS database.

Chinese transactions, the value between 2000 and 2019 was €14.4 billion (Kratz et al., 2020), while the database of Chinese Investment Tracker lists Chinese transactions worth $28.48 billion (American Enterprise Institute, 2020).

Although France has invested much more in China than vice versa, it is increasingly affected by fears of losing its economic sovereignty in the face of a new wave of Russian and Chinese investors, making it a driving force in the screening of FDI in the EU. In the following subsection, we look at the evolution of the legal framework for FDI, then at the French approach to Chinese investment and the sectoral distribution of this investment.

4.2.2 The legal framework

As in all EU member states, FDI by other EU member states cannot in principle be restricted in France, with the exception of the defense sector, where an ex ante notification system has been introduced with rules very similar to those in Germany. France has tightened its FDI screening rules in recent years, but it is the only country where the new measures do not necessarily strike an anti-Chinese tone and also respond to American takeovers. The first case in which a foreign takeover was red-flagged in France was an American takeover, when the French government banned the sale of French Photonis to American Teledyne during the final regulatory approval phase in April 2020 (Andrieux et al., April 9, 2020).

The first law empowering the French Government to adopt and implement specific FDI rules was the French Foreign Exchange Law of 1996, which was amended in December 2014 (Law No. 2004-1343). This version of FDI screening allowed for the monitoring of FDI in specific sectors of the economy. The most recent evolution of the legal framework was Decree No. 2014-479, which expanded the government's authority. In autumn 2018, the French government debated a draft law that proposed to increase the government's scope of action and strengthen the use of so-called golden shares.[8] Under the proposal, companies that do not apply for ex ante approval in strategic sectors could be fined up to 10 percent of the company's annual turnover (Rosemain et al., July 19, 2018). Finally, on November 29,

2018, the government adopted Decree No. 2018-1057, which again expands the scope of FDI screening to the following sectors:

- space operations;
- cybersecurity;
- AI;
- robotics;
- semiconductors and additive manufacturing;
- data hosting;
- systems utilized for capturing computer data or intercepting correspondence;
- IT systems for public authorities in the field of national security;
- information systems utilized in crucial industries; and
- research and development (R&D) of dual-use goods and technologies (UNCTAD, November 19, 2018).

In December 2019, the final piece of France's FDI screening mechanism was created when the government published its Decree (No. 2018-1590), which came into force in April 2020. Two major changes took effect with the new decrees:

1   Biotechnology-related R&D activities have been added to the list of areas requiring approval. It should be added that activities critical to the protection of public health were already covered by previous decrees.
2   Temporarily, the 25 percent investor threshold for non-EU investors was lowered to 10 percent for listed companies, and this temporary lowering of the threshold was valid until the end of 2020. The government argued that takeovers of listed companies with a dispersed ownership structure of less than 25 percent ownership could also be deemed hostile (Barthelémy, 2020).

As we have seen in this subsection, France is strongly focused on FDI regulation. The main question is whether Chinese FDI really targets technology-intensive sectors and thus threatens the core of the French economy. However, before turning to the concrete figures, let us take a look at the arguments put forward in this case.

4.2.3 The French approach to investments and China

The French attitude toward Chinese investment has been very ambiguous in recent years. When it comes to public sentiment toward FDI, it is clear that the trend toward stricter FDI screening rules is part of the bigger picture and is the result of a different economic and foreign policy in France than at the EU level, the aim of which seems to be strategic independence. A tougher stance on China is linked to a more assertive stance on the United States. The French President stressed the sovereignty problems of European nations in October 2020, when he summarized why it is a problem for Europe to rely on American weapons:

"We, some countries more than others, gave up on our strategic independence by depending too much on American weapons systems," ... "We cannot accept to live in a bipolar world made up of the U.S. and China."
(French President Emmanuel Macron, quoted by Momtaz (2020))

At the same time, contradictions in his approach are easy to spot, as national interests sometimes clash with common EU interests, for which the French President also stands. Often hailed as a globalist, Macron clearly wants to strengthen the EU and represent Europe with one voice, which also means a move toward a more protectionist approach to trade and investment. This was also his position on the Belt and Road Initiative (BRI): he argued that the EU should implement a coordinated approach and negotiate with China on the terms of the BRI. At the same time that the Chinese president visited France in 2019, he signed a €30 billion deal with China on the sale of Airbuses. This sharp contradiction between rhetoric and action was pointed out by Peter Koenig:

Yet, surprise-surprise! On President Xi's next stop, Paris, coming from Italy, Macron rolled out the red carpet for the Chinese President and, according to *RT*, went on to sign billions worth of new contracts with the Asian leader. If this looked like a Macron U-turn, it was a Macron U-turn.
(Koenig, March 31, 2019)

This incoherent approach reveals the dilemma of a Common Foreign and Security Policy (CFSP) of the EU, which would function efficiently if applied jointly; but national economic interests often dictate a bilateral approach to relations with China. The economic benefits of cooperation with China are important in the long term. However, countries such as France and Germany have more to lose in this process than Italy, which is much more dependent on external financing, or the United Kingdom, whose economic competitiveness depends heavily on the outcome of the Brexit story. In the case of France, unlike Germany and Italy, there is a particular, albeit less emphasized, conflict with China. Figures show that France's role in the African region has been shrinking for decades: while in 2000 it was the number one exporter to its former colonies, by 2017 it had slipped to third place (Bayes, 2020). Not only has China been prominent in trade (replacing France as number one), but lending to former French colonies has also increased significantly. As of April 2020, 42 of the 52 African countries have joined the BRI! Bayes described the supposed Chinese threat in this way:

> China's emergence in West Africa directly challenges French economic interests. Chinese companies have moved into sectors long dominated by French players: civil engineering, extractives, telecoms, ports. French national champions—Vinci, Eiffage, Orange, Bouygues, Total, Areva, Alstom—must now go toe-to-toe with Beijing's giants, which benefit from the kind of state patronage Paris used to offer its star companies.
>
> (Bayes, 2020)

On the one hand, China can easily leverage the anti-French sentiment in some West African countries. On the other hand, French leaders want to draw attention to the alleged threat of a Chinese debt trap. During a visit to Djibouti in 2019, Emmanuel Macron underlined this threat:

> China is a great world power and has expanded its presence in many countries, especially in Africa, in recent years but what can look good in the short term... can often end up being bad over the medium to long term," ... I wouldn't want a new generation of international investments to encroach on our historical partners' sovereignty or weaken their economies.
>
> (Macron cited by Irish (2020))

After the COVID-19 virus had struck the world in early 2020, French President Macron personally asked China to grant debt relief to African countries. The move was described by some analysts as a publicity stunt, for not only was a "massive cancellation" of debt rejected by the G20, but France's share of the continent's bilateral public debt was 3 percent, while China's share was 20 percent in 2018 (Rivié, 2020).

As we understand it, the balancing efforts of the French foreign policy provide a more rational explanation for the changed and tougher attitude toward China than the alleged interference of Chinese investment with French economic interests.

4.2.4 Chinese investments in French technology-intensive sectors?

Looking at the sectoral distribution of Chinese investment, the energy sector and the tourism sector stand out as the most important target industries. Investments in the energy sector tend to be large investments, the French tourism sector is internationally competitive, but there is only one FDI transaction in the technology sector that could be considered strategic, as it led to the acquisition of French Linxens, which specializes in biometric cards, micro-connectors, RFID, and so on.

In tourism, Accor and Auchan accounted for 38 percent of the transaction value in this sector: 30 percent came from the acquisition of the Louvre

Group and 24 percent from the purchase of Club Med, a French travel and tourism company based in Paris. In the energy sector, only two transactions (2 percent of Total's total assets and 30 percent of GDF Suez's assets) accounted for 91 percent of the total value of Chinese investment in the sector and 21 percent of all Chinese investment in the French economy. The aim of the GDF Suez transaction was to strengthen cooperation on projects (joint investment, financial cooperation, and commercial sponsorship) in the Asia-Pacific region and also in China.

Investments in the technology sector accounted for around 12 percent of all Chinese investments in France. About 70 percent of technology investments consist of a single transaction: the 100 percent takeover of one company's assets (Linxens) by Tsinghua Holdings in 2018 (Wu & Chakravarti, July 25, 2018). That company specializes in the semiconductor industry, but it would be far-fetched to say that Chinese companies would buy strategically important hi-tech firms in France. Interest seems to be much stronger in the energy sector. In France, as in Germany and Italy, investment is very highly concentrated and is targeted at sectors where the country has a long tradition and strong international competitiveness (see Table 4.4).

Looking at the temporal distribution of Chinese direct investment in France over time, the years 2018 and 2019 stand out as good, but at the same time it can be seen that neither the number of transactions nor their volume are significant (see Table 4.5). Moreover, there are years (2009 and 2010) in which the "China Global Investment Tracker" does not register significant Chinese investment transactions in France. Needless to say, we can see the immediate impact of the Global Financial Crisis in 2008 and

Table 4.4 Sectoral distribution of Chinese investment in France between 2005 and 2018[a]

	Million $	*Share (%)*
Energy	6,660	23.38
Tourism	4,810	16.89
Technology	3,590	12.61
Entertainment	3,500	12.29
Transport	2,540	8.92
Other	2,400	8.43
Agriculture	1,650	5.79
Real estate	1,150	4.04
Logistics	870	3.05
Chemicals	700	2.46
Utilities	250	0.88
Health	190	0.67
Metals	170	0.60
Total	28,480	100.00

Source: Own compilation based on American Enterprise Institute's dataset "The China Global Investment Tracker."
a   The data set was updated in early 2020 (American Enterprise Institute, 2020).

Table 4.5 The temporal distribution of Chinese investments in France between 2005 and 2020[a]

Year	The value of transactions (million $)	The share of transactions (%)	The number of transactions
2006	480	1.7	1
2007	700	2.5	1
2008	2,800	9.8	1
2009	0	0.0	0
2010	0	0.0	0
2011	3,750	13.2	2
2012	610	2.1	2
2013	890	3.1	2
2014	3,090	10.8	3
2015	2,820	9.9	4
2016	2,430	8.5	5
2017	1,350	4.7	3
2018	4,840	17.0	8
2019	4,330	15.2	4
2020	390	1.4	2
Total	28,480	100	38

Source: Own compilation based on American Enterprise Institute's dataset "The China Global Investment Tracker."
a   The data set was updated in early 2020 (American Enterprise Institute, 2020).

2019 on Chinese outbound investment. Chinese companies did not invest in Germany in these two years, and there were no investments in Italy in 2008 according to "China Global Investment Tracker" (American Enterprise Institute, 2020).

As we can see, Chinese companies generally do not invest in France for the purpose of technology transfer, but to improve services to Chinese tourists or to improve the market position of their companies (see the GDF Suez case). In our understanding, this is not surprising, given the country's overall performance in innovation and digitalization, which can be characterized by both weaknesses and strengths.

1   In the Bloomberg Innovation Index (2020), France ranked 10th. Comparatively, France did well in business R&D and telecommunications infrastructure, but underperformed in education, and the worst subindex was manufacturing value-added, where the country ranked 39th (Jamrisku, M & Lu, W, January 18, 2020).
2   In the IMD World Competitiveness Ranking, France (32nd) has lost its position from 31st in 2019. Compared to the overall ranking, the country underperforms in infrastructure and AI R&D, while the operating environment in the IMD World Competitiveness Ranking is the second-best among the ranking countries. France performs slightly worse than Germany in the area of AI but is relatively better than Germany in other areas (IMD, 2019).
3   In the 2020 edition of DESI, France ranks 15th out of the 28 EU member states (Germany ranks 12th and Italy 25th in this list). According to DESI,

France is far from the top performers in the EU. The country performs below average in human resources and the use of internet services, but above average in digital public services[9] (European Commission, 2020b).

French firms are far less threatened by takeovers by Chinese firms, especially in the high-tech sector, than German firms. Three French firms (Sanofi, Renault, and Peugeot) are among the top 50 companies in R&D, while eight German firms are on the list (European Commission, 2019: 60). The report of the European Commission (EC) presents the weaknesses of the French economy in this way:

> Most German R&D is in the automotive and industrial engineering sectors, the UK's in pharmaceuticals and software while France has a more diversified R&D sector composition with a much smaller total number of companies than either Germany or the UK.
>
> (European Commission, 2019: 64)

Looking at the group of 2,500 companies listed in the report of the EC, there are 68 French companies whose R&D spending is significant. As for the United States, China, Japan, Germany, and the United Kingdom, it is less surprising that they score better in terms of research intensity, as the size of the market allows for larger companies with larger R&D budgets. However, South Korea and Taiwan have much lower total GDP at both nominal prices and purchasing power parity, and yet their firms' performance in terms of R&D is much better. See Table 4.6, which shows the number of R&D-intensive firms in selected countries.

In sum, France's trade and investment position vis-à-vis China is no different from the average advanced European countries. Its trade is unbalanced and shows a significant French deficit, while the country has invested much more in China than vice versa. So, the argument that the Chinese market is not difficult for French companies to access is not tenable at the macro level, but it is at the micro level, as reports and surveys of European companies consistently show dissatisfaction with Chinese treatment of European companies there. How can China still cope with this situation? The

Table 4.6 The number of R&D intensive firms by country in 2019[a]

US	769
China	507
Japan	318
Germany	130
United Kingdom	127
Taiwan	89
South Korea	70
France	68
Italy	26

Source: European Commission (2019: 27).
a Only those firms are ranked in the list whose R&D expenditures are above €30 million.

price China can pay for its "irregular business environment" is the size of the market that can compensate European companies. One could also say that part of the trade deficit can be explained simply by the fact that it is easier for European companies to serve Chinese consumers out of the market than to export to it. It is, therefore, possible to find a compromise between macro and micro levels, trade, and investment.

Apart from the approach that separates the micro and macro levels, we should add that France has geopolitical interests not only in Western Africa but also in Southeast Asia. There are 8,000 troops in the region, and about one million French citizens live in the Southeast Asian region. Hall summarizes the consequences of this situation for the French position on China:

> But the French leader does not want France or Europe to be hitched to Mr Trump's campaign to decouple the US from China. True to Gaullist tradition, he is a proponent of 'strategic autonomy' for Europe. A new administration in Washington might be willing to work more closely with Europe to counter Chinese unfair trade practices and its push for technological supremacy.
>
> (Hall, 2020)

The desire to achieve a certain degree of "strategic autonomy" can also be confirmed by French foreign policy attempts to reduce tensions with Russia. The French president reached out to Moscow in the summer of 2019 with the reassuring message that Europe will never be stable and secure unless Russia is pacified. Although he did not mention lifting sanctions on Russia, he urged that peace talks with Ukraine be accelerated (Charlton, August 27, 2019). His distancing from the United States is also noticeable: in an interview with *The Economist*, he questioned Washington's commitment to Europe, calling NATO "brain-dead" (*The Economist*, November 7, 2019). This geopolitical approach by the French president shows that he is aware that a long-term alliance between China and Russia would have the potential power to reshape world politics according to their wishes.

4.3 Italy: a case study

4.3.1 Introduction

After the United Kingdom leaving the EU, Italy became the third largest economy in the EU, but the Italian economy is struggling with many more problems than the French or German economies. The almost unsustainable debt that the country has built up from the 1980s (the second oil crisis) to the present day is a heavy burden on the economy despite decades of austerity measures, and unlike Germany and France, the country is far from being at the forefront of technological development, making it more of a market for Chinese goods than an investment destination to access high technology.

Italy is the only country in this group to have joined the BRI. The memorandums of understanding signed by the partners in April 2019 were wide-ranging, covering the banking sector, logistics (ports), agriculture, and construction—50 agreements were signed on this occasion—but this was not the first time that bilateral relations were good. When Romano Prodi, later the President of the EC and Italian Prime Minister, was head of the Italian Institute for Industrial Reconstruction in the 1980s, he ran factory construction projects in China, and later other Italian Prime Ministers also had warm relations with Beijing (Bindi, May 20, 2019). Bindi points out that Berlusconi was the only prime minister in recent decades to have a dissenting voice in the China debate and did not unconditionally support the development of closer relations with Beijing.

The main reason China is showing interest in the Italian economy is not technology, but infrastructure. Fardella and Prodi pointed out that 13 percent of Chinese goods to be sold in the EU pass through the newly acquired Greek port of Piraeus (Fardella & Prodi, 2017), while it was only 2 percent before the port was acquired. The strategic importance of Italian ports can be easily understood from this example.

Bindi lists the alleged threats that would arise from an improvement in relations with China:

1   Bindi warns of the debt trap Italy could fall into if it borrowed from China. We can agree with her in a broader perspective that a growing debt can be detrimental to the future economic development of any country, but in the case of Italy it is clear that the debt has been accumulating for decades and blaming China for Italy's growing debt is a position that is difficult to defend (Bindi, 2019, May 20).

2   She also points out that if relations with China are strengthened, the country could move away from its traditional allies, the United States and other EU members. As far as the United States is concerned, the country has also turned away from its traditional European allies. Therefore, this process has several variables: not only could Italy turn away from the West, but the West, abandoning Italy, could also play a role in this process. Extremely good examples of this were the 2015 migration crisis and the lack of response of other European countries to Italy's hardships in the first months of the COVID-19 crisis (Bindi, May 20, 2019).

In our view, in the case of Italy, there are many more arguments for closer cooperation with China than against it. The main reason why Italy stands to benefit more than it stands to lose by joining the BRI and by engaging in other forms of cooperation lies in the peculiarities of the Italian economy and politics:

1   *Vulnerability to external shocks.* The fundamentals of the Italian economy have not improved significantly since the Global Financial Crisis hit the country. The permanent government crisis combined with the high level of public debt, the traditional North-South cleavage, and the problems in the banking system make Italy extremely vulnerable to

external shocks and could make the country the center of a new European crisis.

2   *Need for capital and technology import.* The country, like the Eastern European countries, needs capital import and new technologies. A country like Italy simply cannot afford to miss out on the growth potential offered by joining the BRI. If there is no other offer that can make a difference in favor of more Western capital and technology import into the country, Italy has to go for China's offer, and right now there is no counteroffer from either the EU or the United States.

3   *A more liberal approach to FDI.* Italy, unlike Germany, has traditionally been a recipient of FDI, so public opinion and policymakers are more willing to accept and acknowledge the need for capital imports. Between 1988 and 2019, according to the World Bank database, there were only 5 years in which FDI outflows exceeded inflows, while German FDI outflows exceeded inflows in 25 years during the same period and in 8 years in the case of France.

4   *Weak performance in new technologies.* As the Italian economy is less specialized in the development of advanced technologies, Italian firms are generally not leaders in this field, and the fear that Chinese firms will steal Italian technology is not widespread among Italian policymakers. In the list of 2,500 R&D-intensive companies published by the EC (European Commission, 2020c: 27), only 26 Italian companies are included. Not only the number of companies, but also their R&D expenditures are disproportionate to the size and development of the economy. Denmark's share is equally high, while the Netherlands, Sweden, and Ireland also surpass Italy in this ranking (see Table 4.7).

5   *The North and the South are divided* on the appropriate economic policies. Italian politicians have recognized that while although there is a need for economic incentives in South Europe, the room for maneuver is minimal, whereas in North Europe there is still room for economic incentives, but economic policymakers do not want to use this tool.

6   *Positive approach for more trade.* As we have seen in the case of Germany and France, unbalanced trade with China is one of the critical points in bilateral relations. Italy's trade with China is similarly unbalanced. Trade volume increased rapidly between 2000 and 2010, but failed to expand further after 2010. At the same time Italy's trade deficit was $20 billion in 2019 and this deficit was much smaller than in 2010 (Zeneli & Capriati, 2020). This improvement is nothing special for the relationship with China, as the total trade deficit, and Italy has a surplus in the trade balance. Italy has realized that if more demand cannot come from the North (because North European countries do not seem to want to expand their aggregate demand in line with German economic policy), Italy will have to look for other markets, and more demand can only be found in expanding markets. The task is not an easy one, as the need

Table 4.7 R&D investment by the 2,500 companies in selected countries (percentage of total spending)[a]

United States	38
China	13.3
Japan	11.7
Germany	10.1
South Korea	3.8
France	3.7
Netherlands	2.3
Sweden	1.2
Ireland	1.1
Italy	0.7
Denmark	0.7

Source: European Commission (2019: 27).
a Only those firms are ranked in the list whose R&D expenditures are above €30 m (European Commission, 2019: 27).

Table 4.8 Italy's trade balance with China ($ billion)

	2010	2018	*The overall balance in 2010*	*The overall balance in 2018*
Italy	−26.7	−20.9	−40.1	46.3

Source: World Bank WITS database.

to rebalance trade relations is relevant, as also pointed out by Italian Minister for Economic Development, Luigi di Maio, who said after the signing ceremony that Italy's aim was to "rebalance an imbalance" in trade (AP News, March 23, 2019). Table 4.8 shows Italy's trade with China.

Despite the good relations between the two countries and the above-mentioned economic reasons for increased cooperation with China, several previously signed deals were never executed, including the announced investment by Huawei ($3.1 billion), which did not materialize because Huawei was excluded from a tender to supply the new generation of 5G technology in 2020. Former Italian diplomat Francesco Sisci explained the sudden turnaround in Italy's approach to China:

The US would consider kicking Italy out of the NATO if Rome uses Huawei's 5G gear. He added that this would place Italy in a dilemma, as a withdrawal from NATO could mean more spending on national defense.

(Yi, 2020)

This threat seems serious enough to change the course of the Italian government, whose economic interests in this case must have been the opposite ones. The next two subchapters examine how Italy introduced and tightened the FDI screening rules. They also examine the sectoral and temporal distribution of Chinese FDI in Italy and raise the question of whether or not geopolitical concerns about Chinese investment can be confirmed.

4.3.2 The legal framework

The general FDI screening mechanism is provided in Italy by Decree No. 21 of March 15, 2012. Scassellati-Sforzolini and Iodice (2019) claim that after six years of application, the law has not deterred foreign companies from investing in Italy. In general, the following sectors are considered strategic: defense and national security, energy, transport, communications, and high technology. These sectors are subject to the prior review procedure mentioned above (Scassellati-Sforzolini et al., 2019: 107–135).

Although the Italian attitude toward FDI was originally more liberal than Germany's, in 2019 the Italian government passed the so-called Decree Law Number 22, which significantly expanded the government's powers; the law came into effect on March 25, 2019, declaring 5G a strategic investment. FDI requires ex ante notification of all contracts and agreements to the design, construction, maintenance, and management of the 5G network if foreign companies (i.e., companies from countries that are not part of the EU) are involved. The government can either prohibit the transaction or require certain conditions from the parties involved (Giarda, 2019).

The EC issued guidelines for the protection of European strategic assets and technologies in early 2020. In response to the guidelines and the COVID-19 pandemic, the Italian government passed a decree (Law Decree No. 23 of 2020) expanding the so-called strategic sector dominated by the "Golden Power," allowing the government to impose restrictions on foreign investment in certain sectors. Foscari et al. summarized the logic and functioning of the "Golden Power" thus:

> Under the Golden Power Law, the Italian government has jurisdiction to review (a) any transaction (i) in the defense and national security sectors, which may harm or constitute a material threat to the Italian government's essential interests; and (ii) in the energy, transportation, communication and high-tech sectors, which may harm or constitute a material threat to the fundamental interests of Italy relating to the security and operation of networks and systems, to the continuity of

supplies and the preservation of high-tech know-how, and (b) solely to the extent that non-EU persons are involved, the execution of any agreement relating to the acquisition of assets or services relating to 5G technology infrastructure (or any 5G technology related components).

(Foscari et al., 2020)

The new law essentially adds AI, robotics, semiconductors, cybersecurity, aerospace, defense, energy storage, quantum, and nuclear technologies as well as nanotechnologies and biotechnologies to the sectors already regulated by the "Golden Power."

4.3.3 Technology-intensive sectors and investments

The hysteria can easily be seen in the case of Chinese investments when news portal headlines suggest changes of epic proportions:

- "China is buying up Italy, one company at a time" (Merelli, October 13, 2014);
- "China swoops in on Italy's power grids and luxury brands" (Sanderson, October 7, 2014);
- "After Pirelli, Chinese shopping spree in Italy to continue" (Aloisi & Arosio, March 23, 2015);
- "How Italy's ruling class has warmed to China investments" (Sanderson & Ghiglione, March 7, 2019); and
- "Selling Italy to Chinese" (*Business Standard*, March 23, 2019).

In contrast to these high-profile headlines, Chinese investment accounted for 1.3 percent of Italian FDI at the end of 2018, according to the Bank of Italy. As this is the most recent data available, we can assume that this figure may have changed in favor of China. Although no more recent data has been published since then, we can safely say that the shares of traditional investors certainly exceed Chinese participation in the Italian economy via direct investment. At the end of 2018, France was the largest investor (16.6 percent) and the US share was 12 percent, while Germany held 9 percent of Italy's FDI shares.

Higher FDI stocks are recorded in the Mercator Institute datasets for China Studies or the China Global Investment Tracker, compiled by the American Enterprise Institute, as they do not include reflows reverse flows to China. Chinese companies invested €15.9 billion between 2000 and 2019, ranking Italy third in the EU (Kratz et al., 2020), the Global Investment Tracker published data on Chinese investment between 2005 and 2019; according to these data, Italy received $26.75 billion in Chinese FDI between 2005 and 2019 (American Enterprise Institute, 2020). Table 4.9 shows the sectoral distribution of Chinese investment in Italy between 2005 and 2019.

Table 4.9 Sectoral distribution of Chinese investment in Italy between 2005 and
2019[a]

	Million $	*Share (%)*
Transport	8,750	32.71
Energy	6,480	24.22
Technology	5,290	19.78
Finance	2,810	10.50
Real estate	870	3.25
Entertainment	840	3.14
Other	790	2.95
Health	720	2.69
Logistic	200	0.75
Total	26,750	100.0

Source: Own compilation based on American Enterprise Institute's dataset "The China
Global Investment Tracker."
a   The data set was updated in early 2020 (American Enterprise Institute, 2020).

1   The transportation sector accounts for about one-third of all Chinese
    investment, and the majority of that investment ($8.75 billion) comes
    from a single investment transaction, the $7.8 billion acquisition of
    Pirelli (American Enterprise Institute, 2020).
2   The second most targeted sector for Chinese investors between 2005
    and 2019 was energy, where two large investments accounted for about
    85 percent of these investments. The value of the two investments, made
    by China's State Grid and SAFE, was $2.7 trillion (American Enter-
    prise Institute, 2020).
3   In the technology sector, Huawei's investments again dominate the
    investment landscape, as the company invested $3.65 billion in Italy
    in four different transactions between 2010 and 2019, none of which
    involved Huawei acquiring the company, but investing in it (American
    Enterprise Institute, 2020).
4   In contrast to Germany, the real estate (3.25 percent) and logistics sec-
    tors (0.75 percent) are underrepresented in the Italian figures, which will
    most likely change after joining the BRI (American Enterprise Insti-
    tute, 2020).

In the case of Italy, it is more difficult to identify patterns or trends in Chinese direct investment. We expect logistics and real estate to play a bigger
role in future data, as the first sector is important due to the country's
geographical location and the second sector could be more important as
the country is a prime tourist destination that can easily attract real estate
investors, although we do not believe that the technology sector will ever be
targeted as much as in the case of Germany. At the same time, we should
add that the value and number of Chinese FDI transactions in Germany
have declined overall. The same trend can be well observed in the EU and

Table 4.10 The temporal distribution of Chinese investment in Italy between 2005 and 2019[a]

Year	The value of transactions (million $)	The share of transactions (%)	The number of transactions
2008	250	0.93	1
2009	0	0.00	0
2010	1,170	4.37	2
2011	130	0.49	1
2012	700	2.62	2
2013	1,300	4.86	1
2014	7,860	29.38	9
2015	10,580	39.55	6
2016	1,710	6.39	5
2017	890	3.33	2
2018	910	3.40	2
2019	1,250	4.67	1
2005–2019	26,750	100	32

Source: Own compilation based on American Enterprise Institute's dataset "The China Global Investment Tracker."

a   The data set was updated in early 2020 (American Enterprise Institute, 2020).

also in Italy, where according to the American Enterprise Institute's China Global Investment Tracker, only one major transaction was carried out. Table 4.10 shows the temporal distribution of Chinese investment in Italy between 2005 and 2019, and also shows that most FDI transactions were carried out in 2014 and 2015: a similar pattern can be observed when looking at the number of transactions.

Le Corre and Sepulchre's (2016) classification of motivations for Chinese firms to invest in Europe contains seven motivations: two geopolitical arguments and five investment strategies (Le Corre & Sepulchre, June 27, 2016). The geopolitical arguments are as follows: Europe is less politicized than the United States and it needs Chinese capital more than the United States.

The investment strategies are:

- the desire to move from cheap products to more sophisticated goods and services;
- the desire to "diversify from the low-margin Chinese market into foreign markets with higher margins";
- the goal of acquiring technology to strengthen their position at home and abroad;
- the goal of better serving Chinese customers in Europe, which is typical of the hospitality industry[10];
- the intention of large state-owned companies (national champions) to expand internationally; and
- take positions of global market leadership.

Based on the Italian characteristics, we have identified six theoretical reasons why Chinese companies could or would invest in the Italian economy. In the list below, we see that only one of these theoretical reasons can lead to a sustained inflow of Chinese FDI, but even this reason may conflict with the long-term interests of the Italian economy.

1 *Italy is less politicized and needs more capital* than other countries. When Le Corre and Sepulcher (2016) drew up their list of motivations, the argument was that Europe was less politicized, but since then the political landscape in Europe has changed profoundly, and perceptions of China have become more blurred in most European countries, as in Italy. For this reason, geopolitical concerns seem to overshadow the originally valid argument of the need for more capital.

2 A logical motivation for entering the Italian market is to *expand the size of the market*, and often, the quantity of goods produced. Italian customers can also be supplied with Chinese goods through trade, but in some cases, confidence in traditional Italian brands can increase sales in Italy. At the same time, this logic often works the opposite way. For example, 75 percent of the famous Ferretti Yacht Group was sold to Shandong Heavy Industry Co. Ltd., which is China's leading manufacturer of bulldozers (Reuters, 2012). Yachts can easily be sold to rich Chinese customers. The acquisition of fashion brands such as Miss Sixty or Cerruti or of the motorcycle-manufacturing arm of Benelli follows the same logic. We argue that the mature Italian market and famous Italian brands will also attract Chinese investment, although this is not always in line with the long-term economic interests of the country.

3 In theory, China could also invest in the logistics sector to accelerate the export of Chinese goods to the Single Market. We see that only a tiny fraction of Chinese investment goes into this sector. Between 2005 and 2019, the logistics sector accounted for 0.75 percent of all Chinese direct investment in Italy. At the same time, we cannot rule out the possibility that joining the BRI could boost Chinese investment in this sector later on (American Enterprise Institute, 2020).

4 *Tech companies* could sell their technology in Italy. Companies specializing in 5G (Huawei, ZTE) would have the greatest potential to sell their products and services as they are ahead of other international competitors, but geopolitical reasons convinced Italian decision-makers not to open this market to Chinese companies. Between 2005 and 2017, these companies accounted for about 17 percent of all Chinese direct investment in Italy, but given the evolving geopolitical backdrop, we do not expect this investment to continue (American Enterprise Institute, 2020).

5 As we have seen in the case of France, Chinese investment does not target Italian high-tech firms to gain access to modern technology. The main reason for this is the country's weak performance in innovation

and digital technologies. According to the Bloomberg Innovation Index ranking, Italy ranks 19th in the world in 2020. The composite index shows weaknesses and strengths in seven innovation dimensions, with Italy scoring best in high-tech density. The IDM World Competitiveness Index shows weak competitiveness in the technological dimension of the composite index. Within the technology subindex, there are other measured indicators, among which contract enforcement, banking and financial services, and the share of hi-tech exports are highlighted as weak factors. Among the strengths of the Italian economy, the total amount of R&D spending, R&D productivity per publication, and robots in education are highlighted as strengths (IMD, 2019: 96–97).

6   The DESI published by the EU summarizes the rather mixed situation regarding new technologies and the preparedness of Italian society:

Italy ranks 25th out of 28 EU Member States in the 2020 edition of the Digital Economy and Society Index (DESI). Data prior to the pandemic shows that the country has a good ranking in terms of 5G preparedness, as all the pioneer bands were assigned, and the first commercial services were launched. There are significant gaps as regards Human Capital. Compared to the EU average, Italy records very low levels of basic and advanced digital skills. ... These gaps in digital skills are reflected in the low use of online services, including digital public services. Only 74% of Italians are regular internet users. Although the country ranks relatively high in its offer of e-government services, public take-up remains low. Similarly, Italian enterprises lag in the use of technologies such as cloud and big data, as well as in the uptake of e-commerce.

(European Commission, 2020c: 3)

As we understand it, recent changes in Italy's regulatory framework regarding FDI from third countries, especially in the area of 5G investment, the recent increasingly negative attitude toward Chinese investment in Italy, and the relatively low level of innovation in the business environment, make the inflow of Chinese investment less likely in the future, but the country's geographical location makes it ideal for logistics investment to have easy access to the Single Market. The reason why Italy is one of the preferred destinations for Chinese companies for direct investment can be explained by its mature market and other traditionally competitive companies whose products and services are attractive to China's emerging and increasingly affluent middle class.

We have seen in this chapter how the regulatory framework for FDI has changed in recent years because of the fear of strategic Chinese (and Russian) direct investment. However, we can conclude that neither the size of FDI nor its geographical distribution explains the geopolitically motivated responses. We also understand that Germany's position is unique among the countries analyzed, as it has the most to lose and gain from Chinese investment in Germany, while France's and Italy's position is somewhat different. Italy, in need of capital and technology transfer, is more open to Chinese business opportunities, while signs of hostility to foreign capital investment in France are sporadic. However, they are not only in response to the challenge from China, but also reflect fear of American and other FDI. In the next chapter, we focus on the Visegrad countries, whose attitudes toward China differ from the French and German reactions.

Notes

1 The ranking includes 39 countries. The following list is the 2019 ranking: Finland, Switzerland, Germany, Denmark, Sweden, the United Kingdom Netherlands, Norway, Luxembourg, Austria, Iceland, Estonia, France, Ireland, Spain, Portugal, Belgium, Latvia, Lithuania, Italy, Slovenia, Russia, Czech Republic, Azerbaijan, Poland, Hungary, Romania, Slovakia, Cyprus, Greece, Bulgaria, Georgia, Serbia, Albania, Republic of Moldova, Ukraine, North Macedonia, and Bosnia-Herzegovina.
2 Military weapons and equipment; specially designed engines or gears; cryptographic systems; dual-use goods or certain military products; operators of critical infrastructure in the areas of energy, water, food, information technology and telecommunications, health, finance and insurance, transport and traffic; developers of software specifically designed for critical infrastructure, operators of telecommunication facilities, and developers of technical implementations for the surveillance of telecommunication; large data centers and cloud computing providers; E-healthcare, license; and media companies (Federal Ministry for Economic Affairs and Energy, 2018).
3 According to AIEI China Global Investment Tracker, between 2005 and 2019 (American Enterprise Institute, 2020), China invested $86.8 billion in the UK. MERICS data show €50.3 billion between 2000 and 2019. Looking at the distribution of these investments, it seems clear that the motivation of Chinese investors is primarily profit, as they invest heavily in less strategic sectors and technology orientation cannot be described as mainstream. At the same time, the traditionally strong sectors of finance and real estate have been targeted by Chinese companies (Kratz et al., 2020). By and large, the investment climate in the UK does not currently appear to be favorable to Chinese investment, although the regulatory framework remains liberal, which does not create sectoral barriers to the entry of foreign investment, particularly technology investment. We can admit that the economic outcome of Brexit cannot be predicted at this stage, and therefore the way in which Britain is *de facto* leaving the EU could significantly alter the incentives for Chinese firms to invest in Britain's technology companies.
4 Based on the average age of the five most valuable firms.
5 Tesla's market capitalization was $347 billion at the end of September 2020, while Germany's largest automaker, Volkswagen, was valued at $92 billion.
6 Balance of payments (BOP) approach data are registered by the ultimate investing country (Bank of France data).
7 It was pointed out in Chapter 2 that the data set of the Mercator Institute for China Studies or the China Global Investment Tracker compiled by the American Enterprise Institute uses a different approach to collect data on Chinese FDI in Europe and other regions. Since they trace the investment back to the

owner and do not include returns to China, these combined annual values of transactions are usually much higher than the data sets with the BOP approach.
8 Golden shares are those shares held by the government which can outvote all other shares under certain circumstances.
9 DESI has five dimensions: human capital, use of internet services, connectivity, integration of digital technology, and digital public services.
10 Chinese companies are investing in Italy with the aim of better serving Chinese customers, especially in the hotel sector, which seems to be a valid reason for Chinese companies to follow this strategy, but the data do not yet confirm this logic, and in the post-COVID-19 period, it may be useful to make major acquisitions in this sector in view of the declining prices of the Chinese economy.

References

Aloisi, S. & Arosio, P. (2015, March 23). After Pirelli, Chinese Shopping Spree in Italy to Continue. *Reuters*, Retrieved from: https://www.reuters.com/article/us-pirelli-m-a-china-shopping/after-pirelli-chinese-shopping-spree-in-italy-to-continue-idUSKBN0MJ20Z20150323

American Enterprise Institute (2020). China Global Investment Tracker. Updated in 2020, *AEI*, Retrieved from: https://www.aei.org/china-global-investment-tracker/

Andrieux, G., Nauges, S. & Ayache, L. (2020, April 9). Recent Decisions Shed Light on Foreign Investments in France. *McDermott Will & Emery*, Retrieved from: https://www.mwe.com/fr/insights/recent-decisions-shed-light-on-foreign-investments-in-france/

AP News (2019, March 23). Italy, China Sign Accord Deepening Economic Ties. Retrieved from: https://apnews.com/article/dae5c72c5cf94e6cb3dce96f9ba9c62f

Barthélemy, P. (2020, May). France Ramps Up Foreign Investment Regulation in the Covid 19 Era. *Jones Day*, Insight, Retrieved from: https://www.jonesday.com/en/insights/2020/05/france-ramps-up-foreign-investment-regulation-in-the-covid19-era

Bayes, T. (2020, May 28). China in Francophone West Africa: A Challenge to Paris. *MERICS*, Retrieved from: https://merics.org/en/analysis/china-francophone-west-africa-challenge-paris

Bindi, F. (2019, May 20). Why Did Italy Embrace the Belt and Road Initiative?. *Carnegie Endowment for International Peace*, Retrieved from: https://carnegieendowment.org/2019/05/20/why-did-italy-embrace-belt-and-road-initiative-pub-79149

Bird, A., Kratz, J., Klärner, H. & Bauer, Y. (2018, November). Tech-Titanen Made in Germany. Eine Perspektive. *McKinsey & Company*, Retrieved from: https://www.mckinsey.de/publikationen/2018-12-05---tech-giants-made-in-germany

Business Standard (2019, March 23). Selling Italy to Chinese. *Business Standard*, Retrieved from: https://www.business-standard.com/article/news-ani/selling-italy-to-chinese-119032300613_1.html

Charlton, A. (2019, August 27). Macron Pushes Outreach to Russia, Offers 'Balancing' Role. *The Associated Press (AP)*, Retrieved from: https://apnews.com/article/4823dda1567d4a618e7f9519e29e81ab

Delcker, J. (2018, July 23). Germany's Falling Behind on Tech, and Merkel Knows It. *Politico*, Retrieved from: https://www.politico.eu/article/germany-falling-behind-china-on-tech-innovation-artificial-intelligence-angela-merkel-knows-it/

Engelstaedter, R. & Gernoth, J. (2014, January). The German Foreign Trade Law and Its Effects on International M&A Transactions. *Paul Hastings*, Retrieved from: https://www.paulhastings.com/docs/default-source/PDFs/stay-current-the-german-foreign-trade-law-and-its-effects-on-international-m-a-transactions.pdf

ESCP (2019). Factbook Digitization. 20 Facts of the Status Quo of Digitization in Germany. *European Center for Digital Competitiveness*, Retrieved from: https://digital-competitiveness.eu/wp-content/uploads/Factbook_digitization_2019.pdf

European Commission (2019). EU R&D Scoreboard. The 2019 EU Industrial R&D Investment Scoreboard, Retrieved from: https://op.europa.eu/en/publication-detail/-/publication/bcbeb233-216c-11ea-95ab-01aa75ed71a1/language-en

European Commission (2020a). Digital Economy and Society Index (DESI) 2020 Germany, Retrieved from: https://ec.europa.eu/digital-single-market/en/digital-economy-and-society-index-desi

European Commission (2020b). The Digital Economy and Society Index (DESI), Retrieved from: https://ec.europa.eu/digital-single-market/en/digital-economy-and-society-index-desi

European Commission (2020c). Digital Economy and Society Index (DESI) 2020 Italy, Retrieved from: https://ec.europa.eu/digital-single-market/en/scoreboard/italy

Fardella, E. & Prodi, G. (2017, September/October). The Belt and Road Initiative Impact on Europe: An Italian Perspective. *China & World Economy, 25*(5), In Special Issue: Eurasian Perspectives on China's Belt and Road Initiative, pp. 125–138.

Federal Ministry for Economic Affairs and Energy (2018). Foreign Trade and Payments Ordinance of August 2, 2013 (Federal Law Gazette [BGBl.] Part I p. 2865), as last amended by Article 1 of the Ordinance of December 19, 2018.

Foscari, F., et al. (2020, April 10). COVID-19-Italy Expands Golden Power Review of Foreign Investments. *White & Case*, Retrieved from: https://www.whitecase.com/publications/alert/covid-19-italy-expands-golden-power-review-foreign-investments

Giarda, R. (2019, April 10). Italy Tightens Foreign Investment Scrutiny Over 5G Technology. *Blog by Baker McKenzie*, Retrieved from: https://www.connectontech.com/2019/04/10/2019-4-10-italy-tightens-foreign-investment-scrutiny-over-5g-technology/

Hall, B. (2020, August 19). Emmanuel Macron's Low Profile on China Is Strategic. *Financial Times*, Retrieved from: https://www.ft.com/content/a132f221-a102-46b6-a81f-635d81a3d4b6

IMD (2019). IMD World Digital Competitiveness Ranking 2019. *IMD World Competitiveness Center*, Retrieved from: https://www.imd.org/wcc/world-competitiveness-center-rankings/world-digital-competitiveness-rankings-2019/

inCITES (2021). Europe 5G Digital Readiness. Retrieved from: https://www.incites.eu/europe-5g-readiness-index

Irish, J. (2018, March 11). Macron warns of Chinese risk to African sovereignty. *Reuters*, https://www.reuters.com/article/us-djibouti-france-idUSKBN1QS2QP

Jamrisku, M. & Lu, W. (2020, January 18). Germany Breaks Korea's Six-Year Streak as Most Innovative Nation. *Bloomberg*, Retrieved from: https://www.bloomberg.com/news/articles/2020-01-18/germany-breaks-korea-s-six-year-streak-as-most-innovative-nation

Koenig, P. (2019, March 31). China—and Macron's U-Turn. *Global Research*, https://www.globalresearch.ca/china-and-macrons-u-turn/5673204

Kratz, A., Huotari, M., Hanemann, T. & Arcesat, R. (2020, April 08). Chinese FDI in Europe: 2019 Update. *Rhodium Group (RHG) and MERICS*, Retrieved from: https://mimderics.org/en/report/chinese-fdi-europe-2019-update

Le Corre, P. & Sepulchre, A. (2016, June 27). China Abroad: The Long March to Europe. *Brookings*, Retrieved from: https://www.brookings.edu/research/china-abroad-the-long-march-to-europe/

Merelli, A. (2014, October 13). China Is Buying Up Italy, One Company at a Time. *Quarts*, Retrieved from: https://qz.com/280247/china-is-buying-up-italy-one-company-at-a-time/

Momtaz, R. (2020, September 29). Macron Calls on Europe to Quit Dependency on US arms. *Politico*, Retrieved from: https://www.politico.eu/article/emmanuel-macron-europe-dependency-us-arms/

Reuters (2012, January 6). Shandong Heavy to Buy Ferretti in $500 Million Deal: Sources. *Reuters*, Retrieved from: https://cn.reuters.com/article/shandong-ferretti-idINL3E8C64SS20120106

Rivié, M. (2020, July 7). Publicity Stunt, Lies and Omission on African Debt, Macron Bets on the Paris Club. Committee for the Abolition of Illegitimate Debt, Retrieved from: https://www.cadtm.org/Publicity-stunt-lies-and-omission-on-African-debt-Macron-bets-on-the-Paris-Club

Rosemain, M., Barzic, G. & Rose, M. (2018, July 19). France to Bolster Anti-takeover Measures Amid Foreign Investment Boom. *Reuters*, Retrieved from: https://www.reuters.com/article/uk-france-investment/france-to-bolster-anti-takeover-measures-amid-foreign-investment-boom-idUKKBN1K92QN

Sanderson, R. (2014, October 7). China Swoops in on Italy's Power Grids and Luxury Brands. *Financial Times*, Retrieved from: https://www.ft.com/content/1bd60160-4496-11e4-bce8-00144feabdc0

Sanderson, R. & Ghiglione, D. (2019, Marc 7). How Italy's Ruling Class Has Warmed to China Investments. *Financial Times*, Retrieved from: https://www.ft.com/content/4b170d34-40f9-11e9-b896-fe36ec32aece

Scassellati-Sforzolini, G., Iodice, F. & Marcon, S. (2019). Italy. In Goldman, S. C. & Koch, M. (Eds.). *The Foreign Investment Regulation Review*, 7th ed. London: Law Business Research Ltd., pp. 107–135.

The Economist (2019, November 7). Emmanuel Macron Warns Europe: NATO Is Becoming Brain-Dead. *The Economist*, Retrieved from: https://www.economist.com/europe/2019/11/07/emmanuel-macron-warns-europe-nato-is-becoming-brain-dead

Tortoise Intelligence (2020). Tortoise Intelligence's Global AI Index, Retrieved from: https://www.tortoisemedia.com/intelligence/global-ai/

UNCTAD (2018, November 19). France Extends Its Foreign Investment Screening. Investment Policy Monitor, *United Nations Conference on Trade and Development (UNCTAD)*, Retrieved from: https://investmentpolicy.unctad.org/investment-policy-monitor/measures/3334/france-france-extends-its-foreign-investment-screening-

World Bank (2020). World Bank WITS Database, Retrieved from: https://wits.worldbank.org/

Wu, K. & Chakravarti, P. (2018, July 25). Chinese Chipmaker Tsinghua Unigroup to Buy France's Linxens for $2.6 Billion. *Reuters*, Retrieved from: https://www.reuters.com/article/us-linxens-m-a-tsinghua-unigroup-idUKKBN1KF0Bl

Yi, D. (2020, August 31). Remarks from French President and Former Italian Diplomat Cast Shadow Over Huawei. *Caixin Global*, Retrieved from: https://www.

caixinglobal.com/2020-08-31/remarks-from-french-president-and-former-italian-diplomat-cast-shadow-over-huawei-101599391.html

Zeneli, V. & Capriati, M. (2020, April 18). Is Italy's Economic Crisis an Opportunity for China? *The Diplomat*, Retrieved from: https://thediplomat.com/2020/04/is-italys-economic-crisis-an-opportunity-for-china/

5 Chinese investment and 5G networks in the Visegrád countries

5.1 The historical background[1]

Most of the Central and European nation-states that emerged or were reborn in the last two centuries resist further integration, mostly because national identity had been forged by the nation- and state-building process of the 19th and 20th centuries. The concepts of "national identity" and "nation-state" cannot be separated from each other in Europe in most cases. This fact has two consequences. Western and Eastern Central Europe together form a more or less coherent region in political, economic, and cultural terms, but Europe has never been unified in a single polity, as it has never been under the control and governance of centralized political power.[2] Probably, this never-ending competition within this fractioned continent was one of the reasons why European countries have contributed a lot to the development of new technologies, solutions, and ideas, starting with the Renaissance and continuing through the Enlightenment, which led the world to the Industrial Revolution and beyond.

As the experience of other regions in the world shows, internal economic competition[3] within Europe, which has had both destructive and creative characteristics, has been a major element of European dominance over the last five centuries. As inward-looking policies and disregard for technological innovation were never real options, they only led to economic backwardness and (asymmetric) dependence.

As the European continent remained one of the core regions of economic development, these countries were all part of an economic division of labor in which resources, ideas, skilled labor, and capital could flow relatively freely between countries at an early stage of historical development. This means that despite the continent's deep fragmentation, a certain degree of integration among these countries was maintained, leaving room for competition.

In this division of labor, Central and Eastern European (CEE) countries lagged behind the West in recent centuries, despite the several attempts that have been made to close the gap between these two parts of Europe. The most remarkable of the concepts used to explain this jigsaw puzzle of

DOI: 10.4324/9781003128625-5

Europe is that of the "two Europe" explanation derived from the work of Leopold von Ranke, who emphasized the importance of the delayed development of this region, which was caused—in his opinion—by the late processes of state- and nation-building (Berendt, 2011: 3–9). Certainly, the process of state- and nation-building in Central Europe started much later than it did in Western Europe.

The paradox with which these countries are now confronted stems from circumstances shaped by globalization and European integration, as both powers would logically push these countries toward closer integration, while the countries of Western Europe seem to accept the concept of an "ever closer Union".[4] Most of the Central and European nation-states that have emerged or been reborn in the last two centuries oppose further integration, mainly because national identity has been shaped by the nation and state-building processes of the 19th and 20th centuries.

In Europe, the concepts of "national identity" and "nation-state" are in most cases inseparably linked. This fact has two consequences:

1   First, some European nations (especially the small ones of Central Europe) are reluctant to engage in further regional integration.
2   Second, in analyzing the EU, we have to accept the existence of different political and economic regimes in the long run. As we understand it, there is no linear convergence of Central and Eastern Europe to Western Europe. Moreover, in different eras, convergence can be reversed and the need to decouple from the West has the possibility of becoming more urgent.

These differences in historical development can be easily observed in the different approaches to economic opportunities (at least) potentially offered by the growing Chinese presence in the region. This seems to be a sudden turn in history, as there was a clear optimism about the future of the region after 1990. Fukuyama's famous phrase "the end of history" accurately characterized the mood of the decade and reflected a kind of "zeitgeist" (Fukuyama, 1992).

Before 2004, there was no question of which political and economic model, neoliberal capitalism, should be followed by these countries, and there was no disagreement about the economic and political benefits of the European project in these countries. Although every new member country benefited from EU membership in terms of EU funds, the question arose: "at what cost?"

Signs of public disappointment could already be felt in some of the new countries before 2008, but the real disillusionment came in 2008 when the financial crisis hit the global economy and hopes of rapid convergence with the West faded. Moreover, in some of the countries, problems of external financing led to a new wave of economic crises in Central and Eastern

Europe, which triggered a renaissance of alternative economic policy and signaled the end of the neoliberal era. It must be made clear that the rise of China has not made the model attractive to the countries of Central and Eastern Europe; only the fact that there is a country that has become successful despite not following either the neoliberal or the Western model, which can lead the CEE countries to rethink their political decisions and experiment more than before with alternative solutions.

5.2 The work of division in Europe

After the Second World War, the countries of Western Europe—not least because of the deepening of the Cold War—transformed their democracies, establishing a more inclusive capitalism in which social democratic ideas together with conservative political forces contributed to the new Europe and laid the foundations for successful future regional integration. In the decades after the Second World War, CEE countries severed their economic and political ties that bound them to the West, but the obvious problems of planned economic development and the failure to catch up with Western Europe made the dependence of the CEE countries on Western European capital and technology clear, even to the naive and true believers of communism in the late 1980s.

After the unsuccessful experiment with communism, most former socialist countries joined the EU in 2004, 2013, and 2017. After 1990, the political and economic transition of these countries followed the model of Western Europe, and these countries copied the political and economic institutions of the West. This so-called "big bang" transition was in sharp contrast to the Chinese transformation, which followed a step-by-step approach and was only economic. The other important difference was that modernization took place through the full opening of these economies to international competition, while at the same time following the political guidelines of the Washington Consensus.

In the 1980s and 1990s, mainstream thinking was dominated by neoliberalism, which offered unilateral and one-size-fits-all solutions. One of the most popular recipes was the Washington Consensus, which originated in 1989, and dominated the 1990s, and the period until 2008–2009. This paradigm, which lost much of its popularity after the Great Recession, was based on two main pillars: more competition and a smaller state (Ostry et al., 2016: 38–41), while in many countries the opening of the economy often led to an externally financed economic growth. This type of growth was fueled by foreign direct investment (FDI) in good times and by foreign credit creating financial bubbles in turbulent times. The recipe of the Washington Consensus also included privatization, open trade policies, and deregulation. After the crisis, this approach fell off its pedestal. One of the main consequences was the end of the growth model dependent on foreign financing.

Another element that has changed economic policy in the post-recession period is the renaissance of industrial policy, which reinforces the idea that competent public authorities are the key to growth.

On the positive side, we can say that the cooperation with Western Europe has created jobs and brought new technologies and capital to the region after 1990, but despite the achievements, the region is still in the semi-periphery when looking at the incomes and other basic indicators of economic development. The biggest problem seems to be that the region's integration into the European framework can hardly increase,[5] suggesting that the potential of this model has been exhausted. The asymmetric dependence or dependency of these countries cannot be cured by further increasing dependence. This is the point at which the interests of Central and Eastern Europe vs. Western Europe diverge significantly.

Both economically and politically, the interests of some countries in the region seem to diverge, although after 1990 the paths of Central and Eastern Europe and Western Europe began to converge again. Despite the profound economic and political transformation after 1990, some of these CEE countries have in recent years begun to turn away from the Western European model. Clear examples are Hungary and Poland, both of which have significant political debates with EU institutions and other Western European countries. We can also point to Albania, Serbia, and North Macedonia, whose disappointment with the EU became extremely palpable when the EU accession process was interrupted by the French president's "no" to EU accession negotiations in 2019 (Delauney, November 2, 2019).[6] Moreover, these countries stressed the lack of any EU solidarity during the first wave of the global pandemic in 2020, when the EU banned the export of medical equipment. The Serbian president expressed his disillusionment this way:

> European solidarity does not exist—that was a fairytale on paper.
> (Serbian President Aleksandar Vuči cited by Çipa, March 20, 2020)

As we understand it, the turning point came in the wake of the Global Financial Crisis, when these countries recognized the following economic problems and put the catch-up process on hold:

1   *Asymmetrical dependence of EU members of the region.* Moreover, the economic dependence on the West did not disappear, as the asymmetry became even greater in the years following 1990. Thomas Piketty pointed out in 2018 that the profits and revenues from real estate leaving the countries of the CEE region represent a much higher number than the EU funds, so the argument that these countries were the clear winners of the EU regional funds is completely wrong (Piketty, 2018).

2   *Asymmetrical dependence of non-EU members of the region.* Excessive dependence characterizes the relations of non-EU members in the

Balkans with the EU. The EU is their largest trading partner, the main destination for outward migration, and the main source of FDI and other capital flows. Moreover, the monetary systems in the region are highly dependent on the euro.

3   *The gap between productivity and wage developments.* Novokmet and Bukowsky emphasize that the gap between productivity and wage development were seen as evidence of asymmetric dependence:

> A rise in productivity is the only way to increase living standards in the long run, which is usually translated to the majority of people through higher real wages. However, average wages have lagged behind the productivity growth in CE Europe, or there was a 'decoupling' between the potential for rising living standards and the actual rise.
>
> (Novokmet & Bukowski cited by Léotard, June 11, 2018)

4   *The problem of foreign ownership.* As productivity gains increasingly went to (foreign) capital owners rather than workers, this also led to large capital outflows.

5   *Challenge for the economic model.* While the foreign sector is characterized by high productivity and is not challenged by the weak bargaining power of the labor force, the domestic sector is littered with small domestic firms and could be characterized by low productivity. Given this scenario, there is a growing risk that this region will remain the low-cost production center of the German economy, specializing in the low value-added segment of global supply chains.

6   *Core–periphery relationship between Western and Central-Eastern Europe.* In the spirit of dependency theories,[7] it can be argued that Central-Eastern European economies are in asymmetric relationships despite the considerable EU funds. The low use of EU funds for domestic research and development (R&D) suggests that this situation will continue.

7   *The end of the road.* Novokmet and Bukowski argue that the convergence model based on foreign capital inflows has not led to the underdevelopment of the Central countries. However, there is a danger that these countries are trapped in a form of specialization in which local value chains are centered around Germany (Léotard, June 11, 2018). The analysis of the National Bank of Poland clearly distinguishes the economic impact of capital inflows, whether in the form of FDI or via the banking sector. Capital inflows via the banking sector made the Central-Eastern European countries more vulnerable to external shocks, while FDI strengthened the manufacturing sectors of these countries. The paper also concludes that this convergence mode has reached its upper limits and that further convergence cannot be achieved in this way (Grela et al., 2017: 88–91).

Not only are many of the basic economic indicators (such as GDP per capita, income levels, etc.) showing how far behind these countries are, but when one looks at R&D spending and other innovation indicators, the problems become clear. On the corporate level, the situation remains unchanged. In the list of 2,500 research-intensive companies, only three Central-Eastern European companies—one each from Hungary, Poland, and Slovenia—made the list (European Commission, 2019: 27). As a reminder: Italy, which has the weakest innovation performance among the three major large economies analyzed, has 26 domestic firms in the elite group of the most innovative companies.

The report Innovation Scoreboard 2020, published by the European Commission (EC) each year, distinguishes four groups of countries: innovation leaders, strong innovators, moderate innovators, and modest innovators.[8] There is no country from the region that has advanced to the group of innovation leaders; only Estonia is rated as a strong innovator, while all other countries in the region are rated as moderate or modest innovators (European Commission, 2020a: 13).

The Digital Economy and Society Index (DESI) of the EC shows a similar state of these countries (see Table 5.1), wherein only Estonia can be characterized as being above the EU average (European Commission, 2020b: 14).

Table 5.1 Basic innovation indicators of the Central and Eastern European countries

	Research and development expenditures in terms of GDP (2018, %))*	*Digital Economy and Society Index ranking***	*Innovation scoreboard 2020 ranking****
EU-27	2.18	–	–
Bulgaria	0.76	28	26
Czech Republic	1.90	17	16
Estonia	1.41	7	11
Croatia	0.97	20	25
Latvia	0.64	18	23
Lithuania	0.94	14	19
Hungary	1.53	21	22
Poland	1.21	23	24
Romania	0.50	26	27
Slovenia	1.95	16	15
Slovakia	0.84	22	21
Montenegro	0.50	–	–
North Macedonia	0.37	–	–
Serbia	0.92	–	–

Source: * Eurostat database; ** European Commission (2020b: 14); *** European Commission (2020a: 13).

5.3 The consequences of backwardness and asymmetrical dependence

The recognition of asymmetric dependence and poor innovation performance led to a search for new policies and out-of-the-box solutions. In several Central European countries, the measures implemented by governments after 2008 reflect a shift toward (re)establishing a state-led capitalist model. This version of capitalism was called "dependent capitalism." Myrant points out to two key factors that distinguish this version of capitalism from other forms:

> ... the level of development of financial systems required for a liberal market economy is absent, as are the cooperative relationships between firms and with trade unions that are at the heart of the notion of a coordinated market economy. These problems are partly overcome with the introduction of a further variety, a dependent market economy, by Nölke and Vliegenthart [2009]. In this version, the CEECs have created environments that give them a competitive advantage in attracting inward FDI by MNCs which then undertake simpler manufacturing tasks in those countries.
>
> (Myrant, 2018: 294)

Hungary and Poland are very clear cases of these problems and efforts to provide policy responses to the new economy. The "re-Polonisation" of the banking sector (Naczyk, 2014: 17) or Hungarian efforts to increase the share of domestically owned banks and financial institutions are examples of this new approach. These countries, after the lessons learned from the Global Financial Crisis (2008–2009), are much more hesitant to continue economic integration within the EU framework. In addition, the introduction of the common currency was taken off the political agenda in both Poland and Hungary. The main argument in this case is that the Euro should be introduced when the country has achieved real convergence (GDP per capita) and not when the nominal Maastricht convergence criteria are met by the given countries. The leader of Poland's ruling conservative party[9] said:

> We say no to the euro, we say no to European prices ... The EU membership treaty doesn't specify the date of euro adoption. Someday we will join, but only when our level of wealth comes close to that of Germany.
>
> (Kaczyński cited by Shah, 2019)

One of the consequences of the political search was the surprisingly strong focus of these countries on China after the 2008–2009 crisis, as China's role grew both as a trading and investment partner and as a source of new technologies in the region. It does not come as a surprise that over the past few years, these countries have tried to diversify their external financing by issuing what have come to be called Panda bonds.[10]

5.4 Cooperation between the CEE countries and China

The policy of opening toward China (see the "Go China" policy of the Tusk government in 2012 or the "Eastern Opening Policy" of the Orban government in 2011[11]) became rather the average in this region, as a significant part of the new global demand came from this region and a deepening of cooperation promised diversification of trade and more capital from China.[12]

It should come as no surprise that in addition to the Belt and Road Initiative (BRI), which these countries have also joined, a new regional cooperation umbrella has been proposed by China and adopted in 2012. The main motivation for setting up the so-called 16+1 initiative was that despite Chinese efforts to strengthen bilateral relations with the region, it seemed easier for China to create a framework for regional cooperation, given the number of Eastern European countries and the relatively small size of their economies,[13] in addition to the network of strategic partnerships signed with several Central European countries (e.g., Poland, the Czech Republic, Hungary, Bulgaria, Romania, Serbia, and Greece).

At the same time, it quickly became apparent that China's "one-size-fits-all" framework has some limitations, which are due to visible fault lines between the 16 CEE countries:

1 Large economies are more able to take advantage of cooperation, while small economies have difficulties cooperating with China. Size also plays a role in bilateral relations.
2 Political identity is also very different in the 17 countries (membership of the EU, Single Market, and the euro area[14]).
3 This new form of cooperation has aroused mistrust in the EU institutions and the EU countries as to the Chinese intentions with this mechanism, pointing to the divide-and-rule tactics conceivably being employed by major powers.

Kavalski offered four reasons to explain the twists and turns in China's CEE relations in 2020 and provide a rationale for souring relations with China. He highlights the next elements:

- the unfulfilled promises of major investment in the region;
- the growing pressure from the EU and the United States during the tech cold war;
- the protests in Hong Kong in the region; and
- the dwindling domestic acceptance of the expensive projects in China in connection to the BRI and the 17+1 cooperation framework (Kavalski, July 30, 2020).

Kavalski points to Bulgaria, Poland, the Czech Republic, and Lithuania, where the mood is changing into a negative or even hostile perception of China. When arguing against China, the focus is very often on human and minority rights issues. The real reason for the change in perception, however, is the much lower direct investment from China than originally assumed. Table 5.2 shows the ratio of Chinese FDI to GDP in these countries. It would be too simplistic to reduce the China policy in these countries to a business decision. However, looking at the countries at the lower end, we can see that these countries have obviously changed their attitudes toward China recently, while countries such as Montenegro, Serbia, and North Macedonia are developing increasingly friendlier relations with China. The Russian threat in the Baltic countries, Poland, and Romania makes it easier for the United States to bring these countries on board with a common platform against China. Countries that, due to their size or for other economic and geopolitical reasons, are more dependent on economic and security support from the United States and China are more willing to accept attempts at geopolitical influence and take sides in the intensifying debates between the United States and China. While China offers the BRI and 17+1 cooperation to these countries, it does not pursue a security policy in the region. For a while, the United States was criticized for offering the North Atlantic Treaty Organization (NATO) only as a security shield for these countries, but the Three Seas Initiative[15] seems to be emerging as a vehicle for American economic support.

In the introduction, we could see that the real motivation for the willingness to cooperate with China within this region is economic rather than geopolitical. We argued that late nation-building, economic backwardness,

Table 5.2 Chinese FDI as percentage of GDP, ranking based on the relative size of Chinese FDI to GDP

	Chinese FDI stock between 2005 and 2020 (billion $)	*GDP (billion $, 2019)*	*Chinese FDI as percentage of GDP*
Montenegro	1.12	5	22.40
Serbia	10.67	51	20.92
North Macedonia	0.65	13	5.00
Slovenia	2.18	54	4.04
Hungary	5.88	161	3.65
Croatia	0.69	60	1.15
Romania	2.11	250	0.84
Bulgaria	0.46	68	0.68
Czech Republic	0.96	246	0.39
Poland	2.28	592	0.39
Latvia	0.10	34	0.29

Source: Own calculation based on World Bank data and the AEI's China Global Investment Tracker (American Enterprise Institute, 2020).

and asymmetric dependence characterize these countries to varying degrees and shape their reactions to the idea of taking further steps toward EU integration.

The diversification of trade and investment relations is an urgent and long-term issue, especially in the Visegrád Four, while there is an urgent need for capital to improve basic infrastructure in the Balkans. The weakest motivation for cooperation with China is to be found in the Baltic countries, where the region's dependence on Western and Northern Europe is so deep that it does not offer any real foreign policy choices, as is the case with the Visegrád countries. Even in this group of countries, the relative importance of Chinese FDI and the leverage of American foreign policy leads to different results. Moreover, it appears that the United States has recently changed course on its foreign policy in the region, and support for the Three Seas Initiative has given a new level to security-oriented NATO cooperation. In the next subchapter, the geopolitical motivation of the countries and their 5G markets will be discussed in three groups: the Visegrád countries, the Baltic countries, and the Balkan countries.

5.5 The Visegrád Four

5.5.1 The Czech Republic

The Visegrád Four include Poland, the Czech Republic, Slovakia, and Hungary. Despite similarities in modern history, there are significant differences in their level of development and their integration into the European and transatlantic framework. Looking at the level of development, the Czech Republic is an outlier in the group, as GDP per capita measured in purchasing power parity is only 8 percentage points below the EU average. GDP per capita was 74 percent in Slovakia and 73 percent in Poland and Hungary compared to the EU average,[16] according to Eurostat data for 2019. This dividing line can be observed in both R&D expenditures and innovation performance. The figures in Table 5.1 confirm this conclusion, as not only is R&D expenditure higher in the Czech Republic, but both the country's Innovation Scoreboard ranking and DESI rankings are more favorable than those of the other three countries.

When it comes to China's image in these countries, MapInfluenCE offers the most comprehensive picture. MapInfluenCE regularly conducts media analyses on how China's image has changed in these countries. In the case of the Czech Republic, they conclude that between 2010 and June 2017, most of the articles analyzed (1,247) were neutral (45 percent), but the rest of the articles were negative (41 percent), while only 14 percent of the media product examined had a positive attitude toward China (Karásková et al., 2018). This result is very much in line with the results of the Pew Research Center's Global Attitudes and Trends survey, in which the Czech Republic was the

Table 5.3 China's perception in the Visegrád countries

	Share of unfavorable views on China* of those who have an opinion of China (%)	Media articles negative about China** (%)	Media articles positive about China** (%)
Czech Republic	57	41	14
Hungary	37	9.4	4.8
Poland	34	3.0	39
Slovakia	48	26	6

Source: * Silver et al. (December 5, 2019); ** Karásková et al. (2018). Data cover the period between 2010 and June 2017. In the case of Poland, data include the next year until June 2018.

country where 57 percent of people who had an opinion on China had a negative attitude toward the Asian country.

Table 5.3 shows the perception of China in the Visegrád Four, although these data do not reflect the increasingly negative perception of China globally; however, it is indicative that the Czech interpretation of China's role had been extremely negative even before the Covid-19 struck or the official visit of Czech Senate President Milos Vystrcil to Taiwan, which triggered a fierce diplomatic dispute between Prague and Beijing (Hutt, September 1, 2020).

Diplomatic relations between China and the Czech Republic have always had their ups and downs. After the political and economic transition in the region, the country hosted the Dalai Lama as well as the Taiwan Premier Lien Chan, after which relations with China became strained. However, with the presidency of Vaclav Klaus (2003–2013) in the early 2000s, who aimed to improve bilateral relations, things began to change. After the President's official trip to China in 2009, which he made in order to reset relations, the real honeymoon began in 2013 when Milos Zeman was elected as Czech President. Kafkades describes this period as follows:

The Czech-Chinese strategic partnership was now being promoted and strengthened in all areas, and not only at the government level. Thanks to the launch of three direct lines between Prague and China (Beijing, Shanghai and Chengdu), the number of Chinese tourists skyrocketed in a few years and reached 620,000 visitors last year—only topped by Germans, Slovaks and Poles. Bilateral forums and common initiatives multiplied, like the 'China-Czech Cooperation Centre under the Belt and Road Initiative' or the 'Czech-Chinese Friendship Association'... hell, even a traditional Chinese medical centre popped up in the city of Hradec Kralove.

(Kafkadesk, February 12, 2019)

The country thus made the turn to China, joining the BRI and the 17+1 cooperation framework. The motivation was to diversify the country's trade and investment relations. However, if we take a look at the economic indicators, it becomes clear that the balance of economic cooperation is far from what was originally hoped. Political relations cooled off in the course of 2020, and the lowest point was reached when the above-mentioned visit of Czech Senate President Miloš Vystrčil to Taiwan took place. The Chinese Foreign Minister Wang Yi told the media:

> Those who attempt to challenge the one-China principle on the Taiwan question are making themselves enemies of the 1.4 billion Chinese people and will have to pay a heavy price for their moves ...
>
> (Wang Yi, cited by CGTN, August 31, 2020)

The Czech trade balance with China is unbalanced—similar to other European countries—and the deficit has been growing steadily in recent years. While the trade deficit was $14.1 billion in 2010, it increased to $23.5 billion in 2018. One thing has not changed, however: in both 2010 and 2018, the relationship with China created the largest deficit. To understand the scale of the problem, we should add that the total trade surplus in 2018 was $17.6 billion, while trade with China created a deficit of $23 billion.[17] This is not unique in the group of Visegrád countries, as the total trade surplus in Slovakia is smaller than the trade deficit with China. See Table 5.4 for a presentation of the Visegrád countries' trade balance with China.

The picture is no better for the Chinese FDI. Table 5.2 already showed the cumulative amount of Chinese FDI between 2005 and 2019 ($0.96 billion) and its relative size compared to the GDP of 2019 based on China Global Investment Tracker (0.39 percent). Figures from the Mercator Institute for China Studies (MERICS) also show similarly low amounts of directly invested Chinese capital: €1 billion between 2000 and 2019 (Kratz et al., 2020). The only source for the sectoral distribution of these funds comes from the China Global Investment Tracker, where five transactions were recorded, three of them in the financial sector, one in the real estate sector, and one in the energy sector.

Table 5.4 Visegrád countries' trade balance with China ($ billion)

	Trade balance with China in 2010	Trade balance with China in 2018	The overall balance in 2010	The overall balance in 2018
Czech Republic	−14.1	−23.5	6.4	17.6
Hungary	−4.6	−4.0	7.3	6.6
Poland	−14.8	−28.4	−17.0	−5.8
Slovakia	−2.7	−3.9	−0.4	0.5

Source: World Bank WITS database.

Given this business and political environment, it should come as no surprise that access to 5G is restricted for Huawei in the Czech Republic. The United States and the Czech Republic issued a joint declaration in 2020 in which the partners agreed to strengthen cooperation on next-generation 5G networks and to conduct a rigorous assessment of suppliers and supply chains. The most interesting part of the statement was that both the United States and the Czech Republic supported the idea that 5G security should be discussed at NATO. The joint statement accompanying the declaration listed similar European efforts to strengthen cybersecurity (US State Department, 2020).[18]

In summary, the Czech Republic is probably the only Visegrád country that is least in need of a further strengthening of relations with China, having concluded that the meager economic benefits of greater cooperation with China do not counterbalance the geopolitical difficulties the country faces if the policy of openness toward China continues. At the same time, it should not be forgotten that this policy also seems to be in line with China's public perception, so it is easy to find public support for it. The same conclusion can be drawn because of the worsening balance of trade with China and the lack of Chinese direct investment in the Czech economy.

5.5.2 Slovakia

Looking at the Visegrád countries, the need to catch up in terms of the level of economic development is very clear in Slovakia, Poland, and Hungary. This is not the case for the Czech Republic, which is very close to the EU average for GDP per capita in purchasing power parity. Despite sharing a common past with the Czech Republic, neighboring Slovakia is much closer to Poland and Hungary in this regard, especially when it comes to the public perception of China: while it is negative, it is not as negative as in the Czech Republic, although it seems to reflect a different reality than that of Poland and Hungary. According to the Pew Research Center's Global Attitudes Survey, 48 percent of Slovak citizens have a negative attitude toward China, and even the views reflected in the media are rather pessimistic about the potential for cooperation with China. According to the data of Karásková et al. (2018), between 2010 and 2018, 26 percent of the media articles analyzed were about China, while the same figure was 3 percent in Poland and 9.8 percent in Hungary. Despite the extensive coverage of China's perception in Slovakia, we must add that Ondriaš is right to point out:

> the increasingly polarized and tribal nature of Slovak politics mean that the attitude of any given citizens towards China can be inferred from his or her self-described political affiliation.
>
> (Ondriaš, 2019: 2)

Among the political powers that see China as a counterweight to liberalism and asymmetrical dependence on Western Europe, a more positive perception can be found, he argues. As we understand it, this phrase or idea can increasingly be applied to Hungary, Poland, or any other CEE country.

However, it is apparent that there has been little success in Slovakia's relations with China. Similar to the Czech Republic and Poland, the balance of trade with China worsened in 2010 and 2018, with the deficit almost doubling. While the data set from the Mercator Institute (Kratz et al., 2010) shows €0.1 billion of Chinese direct investment in the country, the Global China Investment Tracker of the American Enterprise Institute (AEI) simply does not contain any data on Chinese FDI in Slovakia.

Slovakia was one of the first countries[19] to establish relations with China in 2007. As the political transition in Central Europe has reshaped the political landscape, the government adopted the "Strategy of Developing Economic Relations with China for 2017–2020," then an Action Plan, the country also signed the Belt and Road Memorandum of Understanding and joined the 16+1 cooperation framework; but Slovakia is still the least active of the Visegrád Four in terms of relations with China. Slovakia missed the Belt and Road Summits in Suzhou (2016) and in Beijing (2017); moreover, the post of Slovak Ambassador in the People's Republic of China sat empty for a year (2016) (Turcsányi et al., 2018: 380–381). According to Kelemen at al., Slovakia's different approach to China can be explained by two factors, among others: Slovak political elites learned their lesson in the 2000s and believe that structural factors are the key factors in the economic catch-up process, and the realization that opening up the Chinese market would require immense resources (Kelemen et al., 2020: 17).

Given this environment, it should not come as a surprise that the opening of the Slovak 5G market segment to China and any other scientific cooperation with China will be limited and the Slovak 5G market segment will not be opened. Like Poland and the Czech Republic, Slovakia signed a cooperation agreement on high-speed wireless network technology with Bulgaria, Macedonia, and Kosovo in October 2020 (Gramer, 2020).

5.5.3 Poland

While Poland and China signed a strategic partnership agreement in 2011 and Poland launched its "Go China" strategy in 2012, the arrest on spying charges of a regional director of the Chinese telecommunications giant Huawei as well as that of a former agent of the Polish Internal Security Agency marked a sharp end to the honeymoon period between the two countries. The United States and Poland issued a joint declaration in 2019 endorsing the so-called Prague proposals regarding 5G security, so we can make the argument that Poland accepts the political leadership of the United States in this sensitive area.[20]

Even when a few years ago bilateral relations could be described as warm or friendly, the Polish approach to China was less systematic and comprehensive than the Hungarian opening to China. Bachulska argues in this way:

But Poland in fact never endeavored to create a strategic policy like that of Hungary's 'Opening to the East.' Warsaw's attempts to foster political and economic ties with China have instead been fragmented and incoherent, driven by a poor understanding of China and its implications.

(Bachulska, 2019)

He and others argue that Poland's lack of enthusiasm was not long noticed by observers, as it was preparing for the seemingly sudden turnaround in China policy. We can also add that Poland has been placed in the same category or group as Hungary, as both countries are engaged in serious ongoing debates with EU institutions and other member states on the rule of law and other issues that touch the core of each nation's sovereignty. This is probably a factor that has obscured the fundamental difference in China's perception of the two countries.

The Polish turnaround in China policy was not reversed by the effects of the Covid-19 pandemic, argues Ciesielska-Klikowska (Ciesielska-Klikowska, 2020: 4). She notes that transatlantic relations have become stronger during 2020, as illustrated by the visit of the Polish president to Washington in June 2020. She added several explanations for this change: the increase in US troops in Poland, the abolition of the visa application procedure for Polish citizens, and support for Polish energy security and technology development (Ciesielska-Klikowska, 2020: 1).

Another result of the global pandemic is that European countries are increasingly discussing the dangers of their dependence on China. Oertel stresses the opportunities for Poland, saying:

Across the EU, Covid-19 has changed the economic outlook and the assessment of vulnerabilities through economic dependencies in critical sectors within EU member states. The resulting European debate about re-shoring and shortening supply chains poses a challenge to China, but presents an opportunity for Poland and Central and Eastern Europe more broadly. Moving assembly lines to the region and creating clusters of economic activity with improved access to high-quality infrastructure and enhanced connectivity within the region could help countries there move their industries up the value chain.

(Oertel, 2020: 16)

The potential of exploiting the disruption of global supply chains by Poland or any other CEE country is, in our opinion, very limited, as China is not only a production site where cheap labor is available for Western European companies, but it is also an important market, and moreover a market whose importance increased before and will continue to increase after the Covid-19 pandemic.[21] It seems obvious that pulling out manufacturing capacities impeded sales in China. Second, the main reason for the concentration of global manufacturing in China is that China has built up a very strong base with tight and well-functioning clusters of companies, which is very difficult to rebuild in the short and medium term in another country or region.[22] Third, there are also competing regions: Vietnam and Morocco are often mentioned as possible candidates for direct investment by Western European companies (Pandey, May 4, 2020).

Poland differs in its economic development from the Czech Republic, but measured in GDP per capita, we can note many similarities to Hungary. Apart from that, the need to import capital and technology is equally important for both countries. Another common link would be that the picture of China painted by the media in both Poland and Hungary is much less negative than in the Czech Republic and Slovakia, according to the analysis of MapInfluenCE (Karásková et al., 2018); therefore public opinion in both countries is also more favorable toward China than in the Czech Republic, according to the analysis of the Pew Research Center (see Table 5.3).

The economic achievements that stem from Poland's earlier pro-China strategy seem to be poor. Its trade balance with China deteriorated significantly between 2010 and 2018. The increase in the trade deficit with China was 91 percent during this period. The good news for the Polish economy is that the total trade balance in the same period shrank to $5.8 billion (2018), but we should add that the trade deficit with China was almost five times higher than the total trade deficit. The differences between China's share of the Visegrád countries' exports and imports are striking. For both Poland and the Czech Republic, trade is very unbalanced and points to growing imbalances.

The impact of Chinese direct investment on the Polish economy is extremely low, and as a share of GDP, the second-lowest in the country. AIA's Global China Investment Tracker registers $2.28 billion for the period between 2005 and 2020, which is about half the amount invested in Hungary (American Enterprise Institute, 2020). Kratz et al. register an even lower level of Chinese FDI ($1.4 billion) for the period between 2000 and 2019 (Kratz et al., 2020). The Polish Central Bank shows €300 million of FDI from China at the end of 2019, which is only 0.2 percent of the stock of FDI in Poland. Neither Chinese FDI in Poland nor Polish FDI in China is

Table 5.5 China's share of the Visegrád countries' exports and imports (2018, percentage)

	China's import share	*China's export share*
Poland	11.6	0.9
Slovakia	6.0	1.7
Hungary	5.4	1.9
Czech Republic	14.1	1.3

Source: Own compilation based on World Bank WITS database.

statistically significant (Poland invested €163 million in FDI stock in China, which is 0.7 percent of total Polish FDI stock abroad).

In summary, the following factors of Polish disappointment about the opening to China can be summarized:

- the lack of inbound Chinese investments;
- the increasingly growing trade deficits with China;
- the inability of China to contribute to the strengthening of Poland's energy security; and
- China's weak potential to geopolitically counterbalance Russia in the region.

All these elements make Poland more attentive to American arguments when it comes to China strategy. Moreover, if we draw up a similar list of factors motivating Poland to cooperate with the United States, the picture is much more positive. The balance of trade with the United States is also negative, but the deficit is very small, and the US share in the FDI stock is not significant either, sitting at 2.0 percent in 2019 (which is still ten times the Chinese share). More importantly, the implementation of the liquefied natural gas (LNG) terminal in Poland as part of the Three Seas Initiative and military protection against the Russian threat are the trump cards in the hands of the American foreign policy.

5.5.4 Hungary

Relations between the two countries began to develop rapidly after the visit of Hungarian Prime Minister Péter Medgyessy in 2003. Before that, the last official visit of a Hungarian Prime Minister (Ferenc Münnich) took place in 1959! After this sudden start, every Hungarian Prime Minister since Medgyessy has made sure to visit China, and the new government after 2010 continued the policy and strengthened relations with Beijing.

In the previous subchapter, we could see that there are many common elements in the motivations of Poland and Hungary, which is also the reason why these countries were treated similarly in the press and literature for some time. But we can also observe small differences that play a role in the final implementation of their China policy. Hungary's Asia policy is clearer and more strategic. It may be said that the Hungarian initiative came at the right time, as these Chinese initiatives coincided with the "Eastern Opening Policy" that was launched in 2011. The strategy was revised in 2012 with the adoption of a more comprehensive economic growth strategy (the Széll Kálmán Plan). The strategy stresses the importance of diversifying trade and investment. The aim is to double the export of Hungarian small- and medium-sized enterprises to the target regions, with China, Russia, and India being the main partners of these regions. Details of this policy are described by Becsey, who explains that in addition to the establishment of trading houses in emerging countries and the promotion of Hungarian companies (especially small- and medium-sized enterprises), initiatives in the education and tourism sectors are linked to the core of the "Eastern Opening Policy" (Becsey, 2014).

The concept of the "Eastern Opening Policy" is based on the idea that historically, the Hungarian economy has always been dependent on capital and knowledge imported from Western Europe, as we pointed out earlier. This is still the case today. The Great Recession (2008–2009) has clearly revealed the vulnerability of the Hungarian economy, as in 2019, 81 percent of Hungarian exports went to other EU member states, while only 19 percent of exports went to countries outside the EU. The EU's focus on exports can be explained by the dominance of foreign-owned companies, which accounted for 80 percent of all Hungarian international trade in 2017 (KSH, 2018).

Another channel of economic contagion was the use of Western European banks, which was a common element with other Visegrád countries. The subsidiaries of these banks made up the bulk of the Hungarian banking sector, and when they reduced and/or withdrew their loans in the first wave of the economic shock, this triggered a new wave of an economic shock for Hungary. The policy of opening up to the East as a means of reducing the one-sided dependence on Western Europe is thus an economic project of historical significance for Hungary. It is not only a pet project of the current government but also the most important opportunity to achieve an economic breakthrough and go beyond the status of a middle-income country.

Given the background described above, it is not surprising that the specific target indicator of the strategy is the doubling of exports by Hungarian small- and medium-sized enterprises. The strategy does not exclude multinational enterprises, but it does not focus on them either. The main target countries of the strategy are China, Russia, and India, where the potential for trade growth are the greatest.

Another difference with other countries in the region is that the largest Chinese diaspora in Central Europe is in Hungary, whose origin can be traced to the period between 1988 and 1992 when travel between the two countries was not subject to a visa requirement under the 1988 agreement between the two governments. According to the latest official figures, there were 6,800 Chinese citizens with permanent residence in Hungary, but the number of Chinese people must be much higher, as many Chinese have already acquired Hungarian citizenship or were born in Hungary (Irimiás, 2009: 837). It is estimated that the approximate size of the Chinese diaspora can be around 20,000–40,000 people (Mohr et al., 2020: 166).

As shown in Table 5.3, the media's and the public's perception of China is more positive in Hungary than in the Czech Republic or Slovakia, but slightly worse than in Poland. We must emphasize that there is a split between the opinions published in the media and those published in scientific papers. Reports and opinions in the media tend to support the strengthening of relations with China. These reports focus on the progressive elements of the proposal: investment and the jobs created by that investment are often stressed. If a negative tone is adopted in these articles, they are published in newspapers and on websites dominated by the left-wing liberal opposition. In these cases, there are three typical arguments:

1 Why replace one dependence with another?
2 Is this a new way for corrupt politicians to get access to public funds?
3 China is buying up Hungary.

It is clear that Chinese trade surpluses overshadow the goals of the "Eastern Opening Policy," but the Chinese share in Hungarian trade is not yet significant; China's share in Hungarian imports was around 5.4 percent in 2018, while China's share in exports reached 1.9 percent in the same year. In other words, the turn toward Asia is still in its infancy, the trade balance deficit can be improved, and a rejection of the project based on these percentages would be rash, seeing the trade balance with China was improved in the years 2010 and 2018.

A different kind of criticism often serves party interests, since, in reality, it does not focus on China, but on allegedly corrupt Hungarian politicians and high-ranking businessmen who are portrayed as actors who benefit from public investment. One such case involves the railway line linking Budapest and Belgrade, which is to be modernized through Chinese loans. At the same time, the same accusation is repeated when it comes to corruption cases involving EU funds. This negative element of interpretation can therefore be better attributed to Hungarian policy and not necessarily to the perception of China in Hungary.

The argument that "China is buying up Hungary" is flawed, not only in Hungary, but we could see it in other countries as well. China's share of Hungarian inbound FDI stocks was 0.8 percent, while according to the BOP approach data of the Hungarian Central Bank (MNB), only 0.1 percent of Hungarian outbound FDI was invested in China in 2019. Even if taking into account the figures from China Global Investment Tracker ($5.88 billion) (American Enterprise Institute, 2020) or the MERICS data (€2.4 billion), this argument is weak (Kratz et al., 2020).

Unlike other Visegrád countries, Hungary has not signed any deals on 5G networks with the United States, but it has tried to address the security concerns, which were highlighted by the EC. The Hungarian government passed an investment screening law in early 2019. The law is in line with the EU regulation establishing a framework for the screening of FDI that flows into the EU.[23]

When it comes to 5G technology, the Hungarian government does not seem to regard Chinese companies—especially Huawei—as a security threat. Not only did the Hungarian government and Huawei sign a strategic partnership agreement in 2013, but cooperation seems to have become even better since then. Despite the American campaign and claims that the

Chinese tech giant would incur security threats, the Hungarian government is working closely with the company. After negotiations with the regional president of Huawei Technologies, the Hungarian Economy Minister reaffirmed that the Chinese company should help to build the broadband Internet network in Hungary. At the same time, the position of Huawei in Hungary has been weakened by growing American pressure. The Hungarian government has repeatedly emphasized that it regards the development of 5G networks as a business or technology issue rather than a geopolitical one. On the one hand, Huawei investments are important for the Hungarian government. On the other hand, business decisions of the major telecommunications companies are not easily influenced, for example, at the Norwegian multinational telecommunications company Telenor. State ownership is significant at Telenor, but even in this case, the 25 percent ownership share is not enough to allow the state to make strategic decisions.

In summary, the Hungarian approach is pragmatic, as it focuses mainly on the long-term economic benefits of the technology and seems to address the security risks by maintaining a strong role for the state (hybrid model), but at this point, it is not clear whether this government intends to operate the 5G network as an owner or to introduce a hybrid model, which only means that it would play a strong role as a regulator and a law enforcer. Despite this ambiguity, we believe it makes more sense to acquire and claim ownership of the network.

In this chapter, we have seen that the historical development of the region and its particular economic development goals are the main reasons for a different approach to China. But the opening toward—also in the case of Hungary—is not about China, it is basically about diversifying the trade and investment relations of the countries examined. In the same sense, these countries would also need technology imports from China, because it became clear from the analysis that China has an extremely strong position in the telecommunications sector and related technology development, and this is the sector where cooperation would bring the greatest benefit to the Visegrád Four. At the same time, we have also observed that this very segment creates a lot of geopolitical tensions, and cooperation with China negatively affects relations with the United States. Although the Czech Republic's economy is more advanced, the countries analyzed are more or less homogeneous in terms of their economic development and social progress. Despite the limited social and economic homogeneity of the region, the foreign policy responses of the Visegrád Four began to diverge when the United States began to increase pressure on China globally. Hungary is the country that has most consistently adhered to the strategy of balancing. The irony of this situation is that the claim that China is buying up Central Europe cannot be confirmed by data. Moreover, the disappointment with the China policy stems from the lack of investment and technology imports. Looking at the economic realities in the region, the cautious attitude of Chinese investors is understandable, and so it is not surprising that the majority of Chinese investment did not flow into this region, but Western Europe. Hungary is not only an outlier in foreign policy responses, but also in its effectiveness in attracting Chinese direct investment. The country's relative attractiveness to Chinese investors in the region is strong, and this also explains why Hungary is sticking to the rebalancing strategy it initiated in 2010.

Notes

1 Apart from the two small Mediterranean countries—Malta and Cyprus—the EU-11 can be divided into three blocks: the Visegrád Four, the Baltic States, and the Balkan countries, although the real common ground is that they were all once a part of the socialist block. Even the Balkan countries, which are not yet members of the EU, had a socialist planned economy system before the collapse of the Eastern European socialist regimes; but the historical ties go back much further in the development of these countries, which still provides some clues as to why they behave differently or choose different political steps than the Western European countries.
2 There were several attempts to centralize Europe, but neither the Romans nor the Carolingian Empire, the Nazis, nor the Soviets were able to extend their power to the entire continent.
3 Of course, we must not forget that competition was not limited to the development of economy and technology, but also took many forms, including wars.
4 See Article 1 of the Treaty on the EU: "This Treaty marks a new stage in the process of creating an ever closer union among the peoples of Europe, in which decisions are taken as openly as possible and as closely as possible to the citizen" (Consolidated versions of the Treaty on European Union and the Treaty on the Functioning of the European Union, 2012/C 326/01).
5 Obviously this is not the case for the Balkan countries that are not yet members of the EU. The delay in EU membership is also another factor why China seems so attractive for these countries. Karnitschnig summarized the situation in this way:

> With a population of just over 7 million, Serbia, one of Europe's poorest countries, may seem like an unlikely partner in China's drive to play a bigger role on the Continent. But longstanding ties between the two countries, combined with geography, have helped put Belgrade at the center of Beijing's European push. For Serbia and its neighbors, the advantage is obvious: desperately needed investment in infrastructure with few visible strings attached.
> (Karnitschnig, July 13, 2017)

6 In 2020, the EU decided to start accession negotiations with North Macedonia and Albania.
7 Dependency theories first became popular in the 1950s, when early proponents of the idea (such as Paul Prebish, Hans Singer, Celso Furtado, Walter Rodney, etc.) pointed out that the terms of trade of poor countries tended to deteriorate. This thesis supports the concept of the world-systems theory of Immanuel Wallerstein, in which countries are divided into three subgroups: core, periphery, and semi-periphery. Semi-peripheral countries are industrialized with less advanced technology and they cannot control their finances. Looking at the characteristics, most of them can be found in the current Eastern European region.

8 The reports measure the innovation performance of these countries by using 27 indicators along four main types of activities: framework conditions, investments, innovation activities, and impacts.

9 The governing party is called Law and Justice, PiS.

10 Panda bonds are Renminbi-denominated bonds issued by a foreign bank or corporation but sold in China. The market was launched in 2005 with the first issues of panda bonds by the Asian Development Bank and International Finance Corporation. More about the country-specific interpretations of China can be found later in this chapter. The "16+1" mechanism was introduced in 2012 with China and 16 CEE countries. Typically, a summit was held in a different capital every year.

11 The strategy was revised in 2012 with the adoption of a broader growth strategy (the Széll Kálmán Plan). The strategy emphasizes the importance of trade and investment diversification. The aim is to double the export of Hungarian small- and medium-sized enterprises to the target regions, with China, Russia, and India being the main partners of these regions. Details of this policy are described by Zsolt Becsey, who explains that in addition to the establishment of trading houses in emerging markets and the promotion of Hungarian companies, especially small- and medium-sized enterprises, initiatives in the education and tourism sectors are linked to the core of the "Eastern Opening Policy" (Becsey, 2014).

12 More about the country-specific China interpretations can be found later in this chapter.

13 The "16+1" mechanism was introduced in 2012 with China and 16 CEE countries. Typically, a summit was held in a different capital every year: Warsaw, Poland (2012); Bucharest, Romania (2013); Belgrade, Serbia (2014); Suzhou, China (2015), Riga, Latvia (2016); Budapest, Hungary (2017); Sofia, Bulgaria (2018); and Dubrovnik, Croatia (2019). When Greece joined the 16+1 in 2019, the name of the cooperation was changed to 17+1 cooperation.

14 To date, only the Baltic countries, Slovakia, and Slovenia are members of the Eurozone from the region.

15 The Three Seas Initiative goes back to the Polish interwar concept of the "intermarium" (between the seas), whose roots we can find in the medial Polish and Lithuanian Union against Prussia and Russia. The concept was revived in the interwar period and again by the Polish and Croatian presidents in 2015. However, this is where the analogy ends, as the newly launched cooperation focuses exclusively on economic projects, especially energy, transport, and telecommunications networks. The first summit was held in 2016 and 12 countries joined the initiative. With the exception of Austria, all other countries are former socialist countries of Eastern Europe, which have also joined the EU. In addition, one could also say that the participating countries are also members of the 17+1 cooperation framework with China. (There, too, Austria is the exception.) The special feature of the Three Seas Initiative is that both American and Chinese politicians participated in the first Dubrovnik Summit in 2016, but the initiative received stronger political support when American President Donald Trump participated in the next Warsaw Summit in 2017, where the partners agreed to establish a Three Seas Business Forum. Crucial decisions were taken the following year in Bucharest, where a list of collaborative projects was agreed upon by the partners and a letter of intent regarding the establishment of the Three Seas Fund was signed. As for the infrastructure, the "Via Carpatia" highway that would connect Lithuania with Greece and the development of LNG transport (pipelines and sea terminals in Poland and Croatia) were agreed upon. Despite the participation of the Chinese Ministerial Assistant for Foreign Affairs in the first summits, the initiative seems to be rather dominated by the United States.

This becomes very clear if one looks at the financing of the Three Seas Initiative, which, in addition to the Polish and Romanian pledges (€500 million), €850 million was pledged by US Secretary of State Pompeo in February 2020. The funds would be channeled through the United States: International Development Finance Corporation (Atlantic Council, 2020).

16 EU *average* = 100 percent

17 At the same time, we have to be careful when evaluating this data, as even imported Chinese intermediate products could be assembled in the Czech Republic, and exported, or simply imported Chinese final goods could be exported to other EU countries, which would improve the country's overall trade balance.

18 A similar declaration was signed between the United States and Poland in September 2019.

19 Hungarian Prime Minister Péter Medgyessy visited China in 2003 and revived the country's Sino relations.

20 The Prague Proposals are a series of recommendations that were announced at the Prague 5G Security Conference in 2019. The 32 countries participating included Bulgaria, the Czech Republic, Estonia, Hungary, Latvia, Lithuania, Poland, Romania, Slovakia, and Slovenia, which represented the CEE region.

21 The automotive industry is a good example, as after the first wave of the Covid-19 pandemic, the sale of vehicles quickly jumped back in China, while European sales have suffered tremendously during the same period.

22 According to 2019 data, four out of ten Volkswagen cars are sold in China and three out of ten Mercedes and BMW cars are purchased in the Asian country (Pandey, 2020). Leaving the Chinese market would also mean losing the market, as exporting instead of producing locally would be a completely different and far less profitable business model for these companies. (In 2019, 33 Volkswagen plants in China were operating!)

23 The adopted law basically focuses on FDI related to weapons, parts of weapons, munitions, military tools; secret service tools; data processing by a financial institution; areas important for the maintenance of vital societal functions, such as healthcare, the safety of life and property of the citizens; and the provision of economic and social public services. The law attaches high importance to the foreign share of the investment and the related areas.

References

American Enterprise Institute (2020). China Global Investment Tracker. Updated in 2020, *AEI*, Retrieved from: https://www.aei.org/china-global-investment-tracker/

Atlantic Council (2020, February 15). US Commits $1 Billion Dollars to Develop Central European Infrastructure. *Press Release by Atlantic Council*, Retrieved from: https://www.atlanticcouncil.org/news/press-releases/us-commits-1-billion-dollars-to-develop-central-european-infrastructure/

Bachulska, A. (2019, August 13). Trojan Horses, Jilted Suitors. *China Observers in Central and Eastern Europe (CHOICE)*, Retrieved from: https://chinaobservers.eu/trojan-horses-jilted-suitors/

Becsey, Zs. (2014, March). A keleti nyitás súlya a magyar külgazdaságban. *Polgári Szemle, 10* (1–2). Retrieved from: https://polgariszemle.hu/archivum/87-2014-marcius-10-evfolyam-1-2-szam/tudomanyos-muhelyek/594-a-keleti-nyitas-sulya-a-magyar-kuelgazdasagban

Berend, T. I. (2011). Central and Eastern Europe in the World Economy: Past and Prospects. *Hungarian Studies, 2011*(25/2). pp. 216–225, Retrieved from: http://real.mtak.hu/38906/1/hstud.25.2011.2.4.pdf

CGTN (2020, August 31). 'Heavy Price': Wang Yi Issues Warning as Czech Official Visits Taiwan. *CGTN*, Retrieved from: https://news.cgtn.com/news/2020-08-31/Wang-Yi-warns-of-heavy-price-in-challenging-one-China-principle-TozGH3lzdC/index.html

Ciesielska-Klikowska, J. (2020, June). Polish Foreign Policy after the Coronavirus. Poland External Relations Briefing, *China-CEE Institute, 30*(4). pp. 1–4, Retrieved from: https://china-cee.eu/2020/07/08/poland-external-relations-briefing-polish-foreign-policy-after-the-coronavirus/

Çipa, A. (2020, March 20). The EU & The Western Balkans: Love Lost in the Time of Coronavirus. *Emerging Europe*, Retrieved from: https://emerging-europe.com/voices/the-eu-the-western-balkans-love-lost-in-the-time-of-coronavirus/

Consolidated Versions of the Treaty on European Union and the Treaty on the Functioning of the European Union (2012/C 326/01). Retrieved from: https://eur-lex.europa.eu/legal-content/EN/TXT/HTML/?uri=CELEX:12012M/TXT&from=en

Delauney, G. (2019, November 2). European Snub to North Macedonia Fuels Frustration in Balkans. *BBC News*, Retrieved from: https://www.bbc.com/news/world-europe-50260379

European Commission (2019). EU R&D Scoreboard. The 2019 EU Industrial R&D Investment Scoreboard, Retrieved from: https://op.europa.eu/en/publication-detail/-/publication/bcbeb233-216c-11ea-95ab-01aa75ed71a1/language-en

European Commission (2020a). European Innovation Scoreboard 2020, Retrieved from: https://ec.europa.eu/growth/industry/policy/innovation/scoreboards_en

European Commission (2020b). The Digital Economy and Society Index (DESI), Retrieved from: https://ec.europa.eu/digital-single-market/en/digital-economy-and-society-index-desi

Fukuyama,F. (1992). *The End of History and the Last Man*. New York: The Free Press

Gramer, R. (2020, October 27). Trump Turning More Countries in Europe Against Huawei. *Foreign Policy*, Retrieved from: https://foreignpolicy.com/2020/10/27/trump-europe-huawei-china-us-competition-geopolitics-5g-slovakia/

Grela, A., Majchrowska, A., Michałek, T., Muck, J., Stążka-Gawrysiak, A., Tchorek, G. & Wagner, M. (2017). Is Central and Eastern Europe Converging Towards the EU-15? *Narodowy Bank Polski*, NBP Working Paper No. 264 Retrieved from: https://www.nbp.pl/Publikacje/Materialy_I_Studia/264_En.Pdf

Hutt, D. (2020, September 1). Why the Czech Republic Is Baiting China. *Asia Times*, Retrieved from: https://asiatimes.com/2020/09/why-the-czech-republic-is-baiting-china/

Irimiás, A. (2009). Az új kina migráció—a Budapesten élő közösség. *Statisztikai Szemle*, KSH (Hungarian Central Statistical Office) *87*(7–8), p. 837. Retrieved from: https://www.ksh.hu/statszemle_archive/2009/2009_07-08/2009_07-08_828.pdf

Kafkadesk (2019, February 12). Czech Republic, China's Gateway to Europe? (1/3): The Honeymoon. *Kafkadesk*, Retrieved from: https://kafkadesk.org/2019/02/12/czech-republic-chinas-gateway-to-europe-1-3-the-honeymoon/

Karásková, I., Matura, T., Turcsányi, R. Q. & Šimalčík, M. (2018, April). Central Europe for Sale: The Politics of China's Influence. *National Endowment for Policy*, Policy Paper 03, Retrieved from: https://www.amo.cz/wp-content/uploads/2018/04/AMO_central-europe-for-sale-the-politics-of-chinese-influence.pdf

Karnitschnig, M. (2017, July 13). Beijing's Balkan backdoor. *Politico*, Retrieved from: https://www.politico.eu/article/china-serbia-montenegro-europe-investment-trade-beijing-balkan-backdoor/

Kavalski, E. (2020, July 30). How China Lost Central and Eastern Europe. *The Conversation*, Retrieved from: https://theconversation.com/how-china-lost-central-and-eastern-europe-142416

Kelemen, B., et al. (2020, January). Slovakia and China: Challenges to the Future of the Relationship. *Central European Institute of Asian Studies*, Retrieved from: https://ceias.eu/wp-content/uploads/2020/01/Slovakia-and-China-Challenges-to-the-Future-of-the-Relationship.pdf

Kratz, A., Huotari, M., Hanemann, T. & Arcesat, R. (2020, April 08). Chinese FDI in Europe: 2019 Update. *Rhodium Group (RHG) and MERICS*, Retrieved from: https://mimderics.org/en/report/chinese-fdi-europe-2019-update

KSH (2018). Külföldi irányítású vállalkozások Magyarországon, *KSH (Hungarian Central Statistical Office)*, Retrieved from: https://www.ksh.hu/docs/hun/xftp/idoszaki/pdf/kulfleany18.pdf

Léotard, C. (2018, June 11). The Central Europeans, the Bloodsuckers of the European Union? *Political Critique*, Retrieved from: http://politicalcritique.org/cee/2018/the-central-europeans-the-bloodsuckers-of-the-european-union/

Mohr, R., Osváth, G., Sato, N. & Székács, A. (2020). Japán, kínai és koreai üzleti kultúra. *Oriental Business and Innovation Center (OBIC)*. Budapest: Budapest Business School.

Myrant, M. (2018). Dependent Capitalism and the Middle-Income Trap in Europe na East Central Europe. *International Journal of Management and Economic*, *54*(4), pp. 291–303. Retrieved from: https://content.sciendo.com/view/journals/ijme/54/4/article-p291.xml?language=en

Naczyk, M. (2014, July 10–12). Budapest in Warsaw: Central European Business Elites and the Rise of Economic Patriotism Since the Crisis. Presented at the Conference: *SASE 26th Annual Conference.*

Oertel, J. (2020, September). The New China Consensus: How Europe Is Growing Wary of Beijing. *Council on Foreign Relations*, Policy Brief, Retrieved from: https://ecfr.eu/wp-content/uploads/the_new_china_consensus_how_europe_is_growing_wary_of_beijing.pdf

Ondriaš, J. (2019). Slovakia External Relations Briefing: Perception of the 70th Anniversary of the Founding of the PRC in Slovakia. China-*CEE Institute, 21*(4). Retrieved from: https://china-cee.eu/wp-content/uploads/2019/10/2019er0984 (6) Slovakia.pdf

Ostry, J. D. & Loungani, P. & Furceri, D. (2016, June). Neoliberalism: Oversold? *IMF Finance and Development, 56*(2). Retrieved from: https://www.imf.org/external/pubs/ft/fandd/2016/06/pdf/ostry.pdf

Pandey, A. (2020, May 4). China Offers Hope for German Cars after Stimulus Snub. *DW*, Retrieved from: https://www.dw.com/en/china-offers-hope-for-german-cars-after-stimulus-snub/a-53683919

Piketty, T. (2018, January 16). 2018, the Year of Europe. *Le Mond. Blog*, Retrieved from: https://www.lemonde.fr/blog/piketty/2018/01/16/2018-the-year-of-europe/

Shah, S. (2019, April 17). Poland Rules Out Euro Adaption. *Emerging Europe*, Retrieved from: https://emerging-europe.com/news/poland-rules-out-euro-adoption/

Silver, L. & Devlin, K. & Huang, C. (2019, December 5). China's Economic Growth Mostly Welcomed in Emerging Markets, but Neighbors Wary of Its Influence. *Pew*

Research Center, Retrieved from: https://www.pewresearch.org/global/2019/12/05/
chinas-economic-growth-mostly-welcomed-in-emerging-markets-but-neighbors-
wary-of-its-influence/

US State of Department (2020, October 17). The Transatlantic Alliance Goes Clean.
Fact Sheet. *Office of the Spokesperson*, Retrieved from: https://gr.usembassy.gov/
the-transatlantic-alliance-goes-clean/

World Bank (2020). World Bank WITS Database, Retrieved from: https://wits.
worldbank.org/

6 Conclusions

6.1 Technology in the Chinese economic model

The Chinese leadership closed the 5th Plenary Session of the 19th Party Congress in October 2020 by endorsing the 14th Five-Year Plan (2021–2025) and the "Vision 2035" of China. The Five-Year Plan and "Vision 2035" are the most recent strategic documents on Chinese economic development. In the first chapter of this book, when discussing China's previous economic development strategies (the "Made in China 2025" strategy initiated in 2015 and "Strategic Emerging Industries" adopted in 2006), we stressed that the Chinese economic model has several similarities with the "developing state" paradigm, as demonstrated by the example of Japan, South Korea, and Taiwan. We argued that the original model could be characterized by economic development planning with farsighted bureaucracy, huge reservoirs of cheap labor, foreign influence on economic policy, and export orientation. However, we also added that while the Chinese economy has characteristics that make this model unique—the size of the market, the speed of development, and the large regional differences between regions—it also has a strong focus on technological development, which is an inherent feature of the Japanese, Korean, and Taiwanese models as well. The single but fundamental difference is that the size of China and the rapidity of its rise are calling into question the global leadership of the United States, whereas Japan was never a real challenger of American leadership.

In comparison to "Made in China in 2025" and other Five-Year Plans, the emphasis on technology appears to be stronger than ever in the newly adopted "Vision 2035"; but the final details are not yet clear as the detailed version is expected to be passed by the National People's Congress by March 2021. Based on the communiqué issued after the October 2020 summit of party leaders, the creation of "scientific and technological independence and autonomy" and significant investment in technologies such as semiconductors, 5G, and artificial intelligence (AI) appear to be the cornerstones of the strategy. While some analysts stress that innovation and technology-focused growth is just the next logical step in China's economic development

DOI: 10.4324/9781003128625-6

moving up the value-added chains, Prince Michael of Liechtenstein emphasizes other aspects:

> The strategy represents a pivot in economic policy away from exporting manufactured goods. The intention is not to create a Western-style economy, with growth mainly based on increasing consumer demand. Instead, Beijing wants to increase investment in research and development (R&D), while allowing households to maintain their high savings rates.
>
> (Lichtenstein, 2020)

He explains this step with geopolitical motives, saying that the country is preparing for the negative outcome of the great power competition between the United States and China. As we understand it, the two explanations do not exclude each other but reinforce the need to implement the so-called "dual circulation strategy."

The "dual circulation" strategy has become a key priority in the government's 14th Five-Year Plan (2021–2025). The concept of dual circulation was launched by Chinese President Xi Jinping in May 2020. When Xi visited Shenzhen on the occasion of the 40th anniversary of the special economic zone, the "dual circulation" development strategy was once again brought to the center of political discourse. We do not yet know the details, but the emphasis is much more on the domestic rather than the external loop, which could lead to more limited economic activity, including direct investment, worldwide. The move toward self-sufficiency in semiconductors will most likely also lead to a more domestic-oriented innovation strategy.[1]

The move mirrors the attitude and the conclusions of Western countries in the wake of the Covid-19 pandemic. The global pandemic drew the attention of political elites to the vulnerability of global supply chains and external shocks such as a global pandemic.

Discussion about the "dual circulation strategy" is growing among analysts, as it does not mean much to the observer and can be seen as a mere description of the economy in which the domestic economy (the second circulation) is linked to the global economy (the first circulation) (Yao, September 15, 2020). This is the simplest general description of all modern open economies in the world, not a Chinese special case. The China Global Television Network's (CGTN's) guidance on this concept emphasizes that internal circulation refers to domestic economic activities and external circulation refers to China's economic links with the outside world. It also concludes that the concept signals China's intention to reduce the role of international trade in the Chinese economy and also to strengthen its domestic economy (Guppy & Sicong, October 25, 2020).

Yao also stressed that the term "dual circulation" bears a strong resemblance to the term "great international circulation" coined in the 1990s by Deng Xiaoping (Yao, September 15, 2020). The expression is similar, but the content is completely different, as the concept from the 1990s emphasized the export orientation of the economy, while the new version by Xi Jinping attempts to reorient growth toward domestic demand. From this point of view, we must understand two things:

1 Rebalancing the engines of economic growth is not a new idea in China. The need to redesign the strategy of economic development became more pronounced after the Global Financial Crisis in 2009. The idea of creating more strategic independence is not only a Chinese effort now, as other countries reacted in a similar way to the disruption of global supply chains in the wake of Covid-19.
2 The rebalancing includes a shift toward higher added value chains, which is a logical step in the further progress of China's development, and it also reflects the need for a more environmentally friendly way of producing and consuming.

The Covid-19 crisis combined both elements and led to a reorientation of the country's technology development strategy, and it also led to a new world view in Beijing, which sees:

> … the continued decoupling of global supply chains as an enduring trend, and so Beijing now seeks to attempt a new 'big thing'—balancing emphases on both internationalization and self-sufficiency (自力更生)
> …

> (Blanchette & Polk, August 24, 2020)

Blanchette and Polk also stress that the "dual circulation" strategy reminds the observer of supply-side structural reforms, which in turn is again not new in the development of the Chinese economy. The slogan "new normal," which refers to lower growth, has already heralded a new era of economic thinking in China (Blanchette & Polk, August 24, 2020).

Yuzhu argues that the need for change resulted from the loss of comparative advantages in manufacturing. He notes that despite mainstream interpretation, the loss of comparative advantages was not due to rising wages, but to the increase in land prices and property rental costs. He argues that reforms must create a new balance between urban and rural areas, the respective coastal and inland regions (Yuzhu, 2017: 18).

At this point, we can conclude that all the above elements will ultimately reshape economic policy in China and lead to an increased focus on domestic economic trends and security issues, namely:

- the disruption of global supply chains in 2020;
- the growing rift between the United States and China;
- the need to move up the value chain upwards; and
- the rebalancing between regions.

In the EU, the term which has been increasingly used for this situation is "the creation of technological sovereignty." This is the same goal that China would like to achieve in this situation. While Europe is caught between two major powers, and as Kelly puts it:

> The tech sovereignty push is a way for Europe to hedge against an unreliable US embodied by a president openly hostile to the EU, and the rise of Beijing's authoritarian system, now more widely seen as a cause for fear.
>
> (Kelly, September 3, 2020)

China sees itself among powers that fundamentally dislike its political and economic regime, its reaction is the same search for more technological sovereignty.[2] The previous chapters attempted to map China's struggle for technological supremacy with a heavy emphasis on Europe. The next subsections summarize the main conclusion of these chapters.

6.2 The global picture

In Chapter 1, we focused on the narratives of China's rise, on policy reactions to this rise and we compared the technological development of China and the United States. We centered on narratives of China's development as the outcome of its technological competition with the United States and that China is no more dependent on technology development than it is on the narratives surrounding this struggle:

1   Although the concept "China rising" is not helpful to understand the main motivations of China, neither is it adequate to predict the outcome of the struggle for technological supremacy; but the concept shows us the main areas of the competition.
2   In the West, the "decoupling" narrative has gained much influence recently, while China's foreign policy decision-makers stress the benefits of China's economic and political rise for cooperating countries and offer several cooperation frameworks (Belt and Road Initiative [BRI], the 17+1 cooperation framework, Asian Infrastructure Investment Bank, the New Development Bank, etc.). None of the Chinese initiatives are undisputed; however, they do provide new opportunities (at the same time posing threats, too) to emerging Asian and European countries. Despite all the problems created or revealed by these initiatives, these enterprises embrace globalization and growing internationalization, while the American narrative—at least until 2020—has been pushing for the isolation of the Chinese economy. Being forced to take sides in the competition between the United States and China is not an attractive position for European countries to be in, but they might be forced to do so unless the "engagement with China" narrative is—at

least partly—restored in the American foreign policy. Engagement with China offers the chance of peaceful coexistence; however, it fails to reflect the main motivations of Chinese economic and technological development, and thus it is not helpful in predicting China's behavior.

3   We concluded that China's economic development strategy is easier to understand if compared to the "developmental state" model and we assume this is a modernized version of that model. This interpretation in a de-ideologized way shows why China behaves this way in economic development and also that the Chinese economic development goals are not unique: even American criticism resembles those arguments brought up against Japan and Germany at their early stages of development. Other examples of the Asian "developmental state" also demonstrated that there is no necessary link between democratic institutions and the economic success of a country. Therefore, further development of the Chinese economy does not bear the necessity of the formation of democracy. On the contrary, we might face a new version of governance that heavily relies on new technologies.

This is the point where the analogy ends. The effort to achieve technological sovereignty or increase power is not unique to the Chinese model of development. The rise of new technology is already reshaping the boundaries between private and public life in every society, altering the relationship between state and citizens. The price for more security that people have to pay in this new technology environment seems to be a diminishing space of privacy.

Therefore, the EU and European countries not only face the challenge of China and the United States in the area of technological development, but the redefinition of these boundaries requires attention, and it can lead to different responses from these societies. Similarly, European societies struggle with giving a united response to the China challenge and improving international competitiveness in the global technological competition.

In this chapter, we could also see why the EU seems to be bereft of any meaningful participation in the technology competition. Despite the Single Market with around 450 million consumers, the EU's inherently institutionally flawed industrial policy is more or less useless in giving the needed incentives to foster "European champions" of these new technologies.

Any parallels with either the US or Chinese strategies are not present among EU member states. Depending on their historical and economic developmental stage, European countries pursue different policies for managing China's growing economic presence in Europe. The spectrum of policy attitudes ranges from a balancing approach, which tries to exploit the opportunities created by the rise of a new power and the decline of the old one; to a bandwagoning approach, which occurs when a state forges a coalition with the power it perceives to be stronger. Now the question remains, which is the stronger power?

In the second part of Chapter 1, we focused on technological development and looked for signs of Chinese progress and even superiority. We found that China had made great progress in many areas, but the results are outstanding in areas where size matters. There are areas where China is simply leading because of improving social and educational conditions, and these can be exploited due to the immense population of the country which is not only the world's largest, but China has become the world's largest urban nation as well.[3]

The willingness of the Chinese population to embrace new technologies might be a soft factor, but it can be decisive in the long term, along with the rapidly growing purchasing power of the Chinese market. One could see that the domestic market is there to foster "Chinese champions." The strong political support is there, too, to give efficient impulses for technology development within "developmental state" framework. This model offers us a very different approach to technology and economic development, which allows us to conclude that this competition is more of a contest between two economic development paradigms than it is one of competing ideologies.

If there is one area where China has a chance to beat the United States, it is the 5G. It should therefore come as no surprise that it is precisely this sector where the United States is using its huge foreign policy tools to try to stop China's rise. American foreign policy has become particularly active in 2020. The Clear Network Program, which began developing criteria for assessing the trustworthiness of suppliers, was expanded by the Department of State in August 2020. Since then, all EU member states (except Hungary) have introduced regulations to exclude untrusted vendors or committed to Clean Network or the EU 5G Clean Toolbox (US State Department, September 17, 2020).

This does not mean that China would only look at the American efforts to contain China. China's reaction has ranged from "Wolf Warrior Diplomacy," a new hawkish style of Chinese diplomacy, through the signing of the Regional Comprehensive Economic Partnership (RCEP) agreement with 14 other Asian countries to reaching a political deal on the China-EU Comprehensive Investment Agreement at the end of 2020.

While "Wolf Warrior Diplomacy" seems to have backfired strongly and has not impressed European leaders, opening up to the world may trigger positive reactions. As Chinese President Xi Jinping told the media:

> Our aim is to turn the China market into a market for the world, a market shared by all, and a market accessible to all, …
>
> (Xi Jinping, cited by Cheng, 2020)

The quote contradicts the first interpretations of the "dual circulation" strategy, which rather suggests a less open market, a less open attitude to the world. But we cannot really know whether these interpretations are correct, or we just jump to conclusions without knowing the details, which will be public in 2021. Based on the lessons of industrialization (see the Manchurian type and Treaty Port industrialization) and the experiences of "East Asian developmental states," we can be sure that the key element is the cautious balancing between opening and maintaining strategic independence. The EU relationship is a key for China to preserve strategic independence and maintain openness to the world. We can easily use this thought to describe the logical and feasible Chinese strategy for technology-related issues and cooperation.

This chapter not only focuses on the indicators of technological development but also how policy tools have changed over on both countries. Although we tend to center on how the Chinese approach toward the economy, and especially technology development policies, underwent significant changes over time, but the course of American technology development was altered several times too. After the rise of neoliberalism in the 1980s, the United States was successful in many areas of technology development; yet these successes hid the absence of strategic long-term strategy and contributed to a slow erosion of comparative and competitive advantages of the American economy. The growing inequality and the hollowing out of the middle class were just side effects, but effects that later piled up the problems that spilled over into politics, leading to electoral victory of Donald Trump in 2016, who successfully appealed social classes that were on the losing side of this process.

In the American case, the debate is much more dominated by foreign and security policy issues and concerns than by economic development problems, yet the failure to stop China's rise was more of a self-inflicted injury: it was not an inevitable, unstoppable process that could not have been reversed. It is not the product of the flawed American "policy of engagement," but rather the end of the American dream (increasing income inequality, relatively worse position of the American middle class, and growing social injustice) and the lack of comprehensive "enlightened" government reforms that would have prepared the American society and economy to benefit from the new wave of technological changes triggered by the information and telecommunications revolution.

6.3 The EU's China policy

In Chapter 1, we argued that the growing gap between the two major powers—the United States and China—theoretically leaves room for a third actor, the EU. The EU could serve as a bridge between the United States and China, as it is the player with whom nobody is really at odds. Nonetheless, there are certainly more elements that seem to drag the EU toward its old ally, the United States, rather than China.

1 *Security guarantees.* The United States can protect the EU from the Russian threat. China cannot do it. China will not do it. China is an extremely important economic player for the EU, but security issues are still paramount.

2   *Technology factor.* China is a key factor when drafting feasible foreign policies and economic development strategies in European countries. Yet, the EU as a whole seems to be rather reluctant to cooperate more with China, especially in technology-related areas, because it recognizes the need for creating and strengthening technological sovereignty.

3   *Current economic reality.* China's rise is evident in the numbers related to trade and markets regarding the European countries; however its foreign direct investment (FDI) share (both outward and inward stock) is not significant at this moment, and the United States remains a much more important partner for the EU.

4   *Media noise.* Based on media news and articles, we might think that China is a much more important economic partner for the EU than it is. Looking just at the numbers, it is clear that the road is very long and we are far from an asymmetrical relationship with China.

Some factors add to the complexity of forming a coherent China policy in the EU. The economic and geopolitical interests among countries are different, as we showed in Chapters 4 and 5. Speaking with one voice thus encounters problems stemming from the fragmentation of the EU and the complex institutional setting of the EU. Wong argued in 2017 that:

> EU's Manichean view of its options towards China is short-sighted and misleading. At the root of European unease and divisions about China's rise, is the unfinished and unclear issue of Europe's own vision of world order, and the role of Europe in that order. That the European Union is not a single political entity, nor even a coherent supranational entity, but a creature with a mix of supranational and national competencies, is part of the institutional problem which has been discussed by many scholars elsewhere.
>
> (Wong, 2017: 113)

Nonetheless, Europe's problem not only originates from politics, but economic reality also hinders the formulation of an efficient strategy in 5G. We showed that European hi-tech firms lag behind in the international competition, especially in the field of 5G. This weakness can be explained by poor innovation performance in 5G and insufficient funding. But we also argued that market failures could be corrected by an efficient industrial policy of the EU. This problem leads us to the political sphere. There is no industrial policy and the likelihood of establishing a modern industrial policy on the EU level is extremely low, while the other solution—strong and efficient industrial policies on the member states' level—would disintegrate the Single Market. Reliance on American and/or Japanese/Korean technology just postpones a solution to that problem, and it does not give a proper answer to geopolitical problems either, since technological sovereignty of the EU will not be achieved by relying on American or Korean firms. We argued in these chapters that the EU and its members have two options now: either exclude Chinese firms and fall behind in the competition, or choose Chinese, or American, or Korean firms' inclusion in their 5G networks and fail to create technological sovereignty.

In this chapter, it has been shown that because of fundamentally different economic interests, it is extremely complicated to come up with one voice, that is, one single China strategy at the *EU level*. The inherent logic of the Single Market makes the establishment of a powerful industrial strategy impossible at this point, though we do not exclude the possibility that the EU may relax its competition policy and shape its trade and investment policy to be more protectionist than ever before. The current global economic trends (e.g., reshoring, trade war, and restrictions on technology firms and their services and products) just amplify the possibility of this turn. It cannot be seen what policy choices will be taken by the EU; however, we must realize that the debate is not only about China, but the future of the EU too, since finding a coherent strategy demands compromise from every side.

6.4 Major powers: the United States, Germany, and Russia

In chapter titled "Economic and Political Interests of *Major Powers: the United States, Germany and Russia,*" we explored the question of what geopolitical and economic interests the United States, Germany, and Russia have in cooperation or competition with China in Europe, especially in the field of technology and innovation. We chose Germany from the 27 EU countries because the country is best suited to reconcile the different opinions on China, not only because of the size of its economy but also because of its growing cooperation with China. The initiatives such as the BRI and the 17+1 cooperation can move the center of the EU more toward the East of the European continent—not to the East, but toward it—from which Germany could profit most.

We also chose Russia, as the country's response to the BRI is decisive in the initiative's success or failure and it can be a crucial market for hi-tech products from China. Russia and China are in similar positions in many respects: both challenge the existing world order and the global leadership of the United States. Both are sidelined by the United States, which treats them as rivals, the "axis of the weevil" (Mead, 2013). And none of the countries have advanced economies. At the same time, as neighbors they have long-term controversies that are difficult to solve. Aron summarizes the situation this way:

> History and geography militate against an *entente cordiale* between the two Eurasian giants. Authoritarian states sharing a 2,600-mile border, with much of that boundary first imposed by imperial Russia on a weaker neighbor, are hardly ideally set up to build mutual trust.
>
> (Aron, April 4, 2019)

Yet, Russian and Chinese cooperation in technology development can be decisive in the outcomes of the technological race between the United States and China.

The chapter first focused on the record of the United States' efforts to turn European countries against China, which has been mixed. We concluded that the foreign policy of the United States is no different in the EU than elsewhere: it pursues global goals in different locations, in this case, the containment of China. The EU, and especially Germany, has a special role in this policy, because if they can be persuaded that the alleged threats that China poses to the European continent and to the rest of the world are real concerns, then this would certainly guide Asian and African countries in their policies, too.

As Chinese FDIs continue in the EU, the "thirst for technology" and concerns about intellectual property rights have sparked a fierce debate over the growing Chinese presence in the EU. We focused on these elements in this chapter.

1   *Chinese FDI in Europe.* It could be concluded from the figures presented that the general fear in connection with Chinese investments cannot be justified given the relative insignificance of Chinese investments in the EU and *Germany.* In China, European companies are indeed more restricted in their daily business activities than Chinese companies in the EU, but the size of the Chinese economy can compensate European companies and countries for this loss. For example, the German FDI stock in China is much higher than the Chinese FDI stock in Germany, so the reciprocity that Germany has always demanded has been achieved at the macro level. In Russia, the problem is not the strong inflow of Chinese direct investment, but the lack of it. The Russian position is very similar to the position of many Central European nations, as these countries have a common need for more capital and technology.

2   *China's growing thirst for technology* puts Germany in the most threatened position in this cooperation, especially the German automotive industry, which is in a position where losses and profits can be particularly high. Under this aspect, different groups of EU members can be distinguished. Scandinavian countries can indeed be targets for Chinese companies, while the Visegrád Four would like to have more access to Chinese technology. Central European firms are less likely to be targeted by Chinese firms as the innovation potential to be tapped is low or the technology Chinese could have access to is scarce.

3   Regarding the *debate on intellectual property rights* between the EU and China, we concluded that although "theft" is not ethical, it has been an integral part of the development of other emerging economies in the past, and China follows their pattern of behavior through reverse engineering, that is, China will only advocate adequate protection of rights if it loses more than it can profit from this behavior.

Trade deficits characterize the relations of all actors with China (the United States, Germany, and Russia), which in our understanding is simply due to the fact that China is the manufacturing hub of the world, while the United States and Germany are specialized in high value services and Russia in natural resources. American reshoring efforts to change this situation have become more apparent after 2016 under the Trump administration. The goals of reshaping global supply chains are unlikely to fade under the Biden administration, but the strategy of "reshoring" also implies a change in trade routes and volumes and also affects technology transfer between countries. It can therefore be assumed that reshaping global supply chains is also part of the technological competition between China and the United States. Looking at the reactions of European countries, reshoring can lead to "nearshoring" at the maximum, and reallocating manufacturing capacities to low-cost Central European countries or North Africa.

We also analyzed the Russian motives for cooperation with China. We emphasized that Russian trade and investment relations with China are very unbalanced and that long-smoldering conflicts of interest could change the friendly relations between the two nations easily. We must also add that the two partners have very different objectives; Russia, like the Visegrád countries, focuses on FDI and technology inputs, while China's main interest is in expanding its markets and securing much-needed natural resources. As we understand it, the logic of "the enemy of my enemy is my friend" can only function temporarily, and with the rise of China and possibly the Chinese influence in Russia, the motivation to make "new friends" and to counterbalance China's influence in Russia will increase slowly in the long run. As we understand it, this logic cannot dynamically stimulate these relations, but it does provide a solid basis for moderate growth and future cooperation at the moment, where Russia will be the one to slow down the process.

6.5 China polices of the three old members of the EU

We have explicitly addressed the different interests regarding China policy in the chapter on "Chinese Investments and 5G networks in Western Europe." The three main EU destinations for Chinese FDI are Germany, France, and Italy, but the share of Chinese investment in Germany, France, and Italy is very low[4] compared to other sources of FDI (such as the United States, Canada, Switzerland, or Norway). At the same time, France and Germany have invested significantly more in China than vice versa, and Italy needs more capital imports, making it the only country in this group that has received more FDI from China[5] than it has invested in the Asian country.[6] As far as the patterns of Chinese FDI investment are concerned, only in Germany can the geopolitical fear of being targeted by Chinese firms to gain access to technology be confirmed. Both the French and Italian acquisitions can be considered "traditional" investments that focus on the traditional strengths and sectors of these economies.

We should point out that the more advanced countries of the EU are more likely to be targeted by Chinese FDI than the less developed ones. This is also true for the Netherlands, which according to the Mercator Institute's database (Kratz et al., 2020) received around €10 billion in Chinese FDI between 2000 and 2019 and $15.9 billion between 2015 and 2019 according to the American Enterprise Institute (AEI) China Global Investment Tracker. The country is the fourth most important EU country for Chinese direct investment. Table 6.1 shows the sectoral distribution of Chinese investment, although it can be observed that investment in the technology sector is much more important than in other sectors. At the same time, these investments can be irregular in terms of their sectoral distribution, the best example being Finland, where Chinese firms invested $15.6 billion, but 90 percent of FDI was two acquisitions in the entertainment sector. To sum it up, Chinese investment in Europe seem more motivated by profits than long-term geopolitical considerations.

Looking at the legal frameworks in the three countries analyzed, Germany offers Chinese investors the toughest legal framework and Chinese investments face the greatest challenges here, although it should be emphasized that it is here that the greatest strategic advantages of the investments can be achieved, as the acquired companies are innovation leaders in the transport and technology sector and highly competitive on the international market. The fact that the German Chancellor maintains regular contact with Chinese decision-makers is positive and confirms the practical approach of German policy; however, the benefits of this cooperation could become clear to the German leadership if trade between the two countries becomes more balanced.

Table 6.1 Sectoral distribution of Chinese investment in the Netherlands between 2005 and 2019[a]

Technology	6,690	41.94
Agriculture	4,020	25.20
Finance	1,770	11.10
Transport	830	5.20
Logistics	610	3.82
Real estate	560	3.51
Metals	470	2.95
Tourism	390	2.45
Health	270	1.69
Chemicals	230	1.44
Other	110	0.69
Total	15,950	100.00

Source: Own compilation based on AEI's dataset "The China Global Investment Tracker."
a The dataset was updated in early 2020 (American Enterprise Institute, 2020).

In France, the picture is very similar to the economic impact of Chinese investments, but the policy approach is very different. The confrontational style of the French President creates a rather hostile environment, and at the same time, the rhetoric underlining European values and a concerted European attitude toward the Chinese is sometimes in sharp contrast to French actions. The French approach can be very lenient when China seems to be willing to buy French products, as seen in the example of Airbus. It also shows what negotiating strategy should be followed by the Chinese to achieve results. The French case is the only one of the three countries analyzed in which hostility is directed against foreign investors in general, as an anti-American tone is as typical in these debates as the comments against Chinese investment.

Like Germany and France, Italy strengthened FDI screening mechanisms and adopted several amendments in 2020, including explicit 5G. As Italy needs more capital and better technology, it is obviously the country that stands to benefit the most from cooperation with China under the BRI framework. However, we have identified only two long-term motivations for Chinese companies to invest in Italy: (1) investment in the logistics sector and infrastructure can create faster trade routes for Chinese goods to the Single Market and (2) acquisitions of Italian brands in traditional sectors primarily serve the needs of affluent Chinese middle- and upper-class customers to gain access to luxury goods and services at lower prices and to allow investing Chinese companies to enter the European market.

The legal frameworks in the countries studied have changed from a more liberal to a more differentiated approach, which can be considered more appropriate to their economic development goals and national interests; however, it is questionable whether strategic decisions are made without an ideological bias and with reference to national interests. In our opinion, the next key problems can be identified from this perspective:

1 European hi-tech firms are lagging behind in international competition, especially in the area of 5G.
2 This weakness can be explained by poor innovation performance in 5G and insufficient funding.
3 The above problems could be corrected by an efficient industrial policy of the EU, which does not exist at present. The likelihood of establishing a modern industrial policy at the EU level is low and a strong industrial policy at the member state level would lead to the disintegration of the Single Market, which again can be ruled out. In other words, we can forecast the continuation of these problems.
4 The involvement of American or Korean companies does not solve the problem, but only postpones the solution and does not provide an adequate response to the geopolitical problems, because the technological sovereignty of the EU cannot be achieved by relying on American or Korean companies.

5   At this point, the EU and its members have two options: either exclude
Chinese companies and fall behind in competition or decide to include
Chinese, American, or Korean companies in the 5G networks and give
up on "technological sovereignty."

Another bias, often repeated by analysts, is that Chinese FDI to Europe is
mainly motivated by geopolitical factors, and direct evidence of this is the
large proportion of state-owned companies among Chinese firms investing.
The latest data from Mercator Institute for China Studies (MERICS) show
that this argument might have been valid a few years ago, but this ratio of
state-owned companies within Chinese firms investing directly in the EU
has decreased significantly in recent years. Between 2010 and 2015, this ratio
was 70 percent, falling to 11 percent in 2019 (Kratz et al., 2020). Kratz et al.
argue that the declining share of state-owned enterprises can be explained
by changes in EU legislation, global trends, and China's economic restrictions. The last argument—maybe the most important one—is that China's
newly launched dual circulation policy will most likely accelerate the trend
of falling Chinese FDI.

We should add that when China's economic vows explain the decline in
both the share of state-owned enterprises and the total value of Chinese
FDI in Europe and the world, this indirectly proves that Chinese investment is mainly motivated by economic interests and far less politically (or
geopolitically) driven than others suggest. The sectoral distribution of Chinese FDI in the EU also supports this argument, although Information and
Communication Technology takeovers accounted for 20 percent of total
Chinese investment in the EU, 40 percent in 2019 went to consumer goods
and services (Kratz et al., 2020)

Trade between China and the three countries analyzed is unbalanced:
Germany was China's most important trading partner for the fourth year
in a row in 2019 and its trade with China is very close to being balanced. In
our view, Germany's long-term economic interests will outweigh the foreign
policy efforts of the United States to find a unified Western front against
China. Stern expressed this disappointment in this way:

But on any issue impacting Germany's economic well-being, Berlin's
actual decision-making is remarkably consistent. In addition to securing ties with China, Merkel is currently defending the Nord Stream 2
gas pipeline with Russia against threatened sanctions from the U.S.
Senate, and low levels of defense spending against White House plans
to withdraw 12,000 U.S. troops from German soil. As far as Berlin is
concerned, Americans can list the sacrifices they've made for German
security and prosperity until they're blue in the face. The benefits to
domestic stability of economic cooperation with strategic rivals remain
a core German national interest.

(Stern, October 13, 2020)

At the same time, France and Italy's trade relations with China are more problematic because the chances of rebalancing the imbalance seem very slim now. In our view, France's tough tone in relations with China is guaranteed, given the geopolitical conflicts of interest mentioned above, while Italy with a "realpolitik" approach can either deepen relations with China or adopt a tougher stance toward China, depending on Italy's economic interests. At the moment the BRI has tipped the scales in China's favor, but if China's promises are not kept or a counteroffer from the EU or the United States arrives, Italian opinions can change quickly.

From a broader perspective, we can argue that less globalization—decoupling China or any other country from the world economy—would harm global growth in the medium and long term, and therefore not improving economic relations with China would be a strategic failure. Frank Pieke argues:

> Europe needs to disentangle itself from this spiral of aggression driven by binary, winner-takes-all perspectives. As it does not aspire to be a superpower, Europe can deal with Beijing with more nuance than the US—China is indeed a threat in some areas but remains a positive force in others. This is not an economic or a military challenge—it is a political one. How does Europe decide what to share and withhold? It needs to answer that question—not isolate China.
>
> (Pieke, August 27, 2019)

To sum it up, we found out that there are differences not only between the West and the East of the EU, but also between the analyzed Western European countries like France, Germany, and Italy:

1 As far as the patterns of Chinese FDI investment are concerned, it is only Germany where the geopolitical fear of being targeted by Chinese companies to gain access to technology can be justified. Both the French and Italian acquisitions can be considered "traditional" investments that focus on the traditional strengths and sectors of these economies.

2 Italy is the only country in the group that is in such urgent need of capital and technology imports, so it is also the country that could benefit most from cooperation with China under the BRI framework. Nevertheless, we have identified only two long-term motivations for Chinese companies to invest in Italy: investments in the logistics sector and infrastructure can give Chinese companies faster access to the Single Market and acquisitions of traditional Italian brands.

3 France's position is unique, in that it has a geopolitical conflict of interest with China in Africa, and like Italy the country has not been able to rebalance its trade with China. So, France has two powerful arguments for holding a more hostile attitude toward China (see Table 6.2).

Table 6.2 Characteristics of Chinese FDI and the legal framework in France, Germany, and Italy

	FDI screening adopted?	FDI screening's legal framework changes recently? When?	Any discernible pattern in Chinese investment?	The two main targeted sectors	The aggregate share of the targeted sectors within the Chinese direct investment (%)	The aggregate value of Chinese investments in the countries between 2005 and 2018 (billion US$)[a]	China's share in inward FDI stock of the given country (%)[b]	Invested more in China than vice versa	In which aspect is the country different from the other two analyzed countries?
France	Yes	Yes, 2019	Yes	Energy, Tourism	40.3	28.48	1.2	Yes	Conflict of geopolitical interests with China
Germany	Yes	No, 2020	Yes	Transport, real estate	59.4	47.58	0.9	Yes	Hi-tech firms can be targeted by Chinese acquisitions
Italy	Yes	Yes, 2020	No	Energy, transport	56.9	26.75	1.3	No	Italy invested less in China than vice versa

Source: Own compilation.
a Based on AEI's dataset "The China Global Investment Tracker." The dataset was updated in early 2020.
b Balance of Payment approach data; in all cases data are registered by the ultimate investing country (Bank of Italy, Bank of France, German Federal Bank).

In our opinion, Germany is best placed to reconcile different opinions and to strike a balance between foreign policy aspects and economic interests in the EU. The development and size of its economy and its attractiveness for Chinese investors enable Germany to be a leading and decisive voice on the China issues—yet this can only be done in agreement with France. The potential to benefit from technology cooperation with China is greatest in the case of Germany. Germany has not automatically excluded Huawei from the market, but it would be very difficult for Huawei or other Chinese firms to enter this market segment. Apart from a strongly media-focused area, the potential for cooperation—despite threats—is great in the manufacturing sector, especially in the automotive industry.

6.6 China policy in the Visegrád countries

In the chapter on "Chinese Investment and 5G Networks in the Visegrád Countries," we began with a historical introduction that highlighted the differences in economic interests between the West and Central Europe. As far as capital and technology in the economies are concerned, the Visegrád countries are in an asymmetric dependence on Western Europe, and this dependence results from a long historical development which, for this reason, is not easy to change. It follows that policy responses in these countries are different from those of Western European ones. This discrepancy creates problems in speaking with one voice with respect to China, but at the same time we have also seen in this chapter that the China strategy of these countries is more about diversifying the investment and trade relations of the respective countries than about China itself.

When the Global Financial Crisis hit Europe in 2008 and 2009, Western European countries chose their own path and focused on alleviating the suffering caused by the crisis at the member state level, which is why it was so difficult and time-consuming to find a response to the Greek crisis. The lesson that the Visegrád countries drew from this crisis was that the growth model (dependent market economy or convergence model) that these countries implemented after 1990 reached its upper limit, and new ways must be found to accelerate the process of catching up with the West. The China policy must be interpreted in this framework, and therefore it can be argued that the Visegrád countries will continue their economic cooperation with China in the future if conditions are favorable; but we can also argue that they will avoid conflicts with the United States over their relations with China for two reasons. The benefits of diversification may also arise from other relationships (Japan, India, South Korea). The example of Hungary shows that China is definitely not the number one investor from the Asian region; other Asian countries can fill this vacuum. And unbalanced trade is an important factor that can put relations with China on the backburner. This has happened in the Visegrád countries with the exception of Hungary.

It is extremely painful for the economic development of these countries that technology transfer has become the focus of the intensifying geopolitical debate between the United States and China, as this is the element of economic cooperation with China that these countries would most need. After years of intensified cooperation under the BRI (2013) and 17+1 cooperation (2012), hopes for intensive capital and technology import have faded. Relative success can be seen in the example of Hungary, which not only attracted the most Chinese FDI as a percentage of GDP but was also successful in attracting investment from Huawei.[7]

We also said that these results are significant in regional comparison, but they are not when comparing the value of investments or their impact on employment with advanced Western countries. So, we can conclude that media coverage of China buying up Europe, especially Central Europe, is heavily biased and influenced by geopolitical prejudices. If the Visegrád countries can criticize Beijing's behavior, it is not for its overinvestment in the region, but for the lack of it. Hutt and Turcsányi use the phrase "excess of attention":

The excess of attention on the region's allegedly growing economic dependency on China has overshadowed the reality, in which a few well-positioned politicians and businesspeople have hijacked the relationship for personal gain. In truth, Central and Eastern Europe is far less economically dependent on China than perhaps any other region in the world—and especially compared to the rest of the European Union.

(Hutt & Turcsányi, May 27, 2020)

The problem is rather that Chinese investment bypasses Visegrád countries and goes to wealthy Western European countries, which is indirect evidence that Chinese investment is mainly motivated by profit rather than geopolitical thinking. Moreover, if one looks at China's investment and trade positions in the EU countries, one can argue that these countries are not dependent on China, or much less than the Western European countries (Hutt & Turcsányi, 2020, May 27). The argument can thus be reversed, and it is not Central Europe which undermines the EU, but Western Europe. However, this line of argument is far removed, since Chinese investments in Europe (as we have seen in the previous chapters) are mainly motivated by profit and less so for geopolitical reasons. Hutt and Turcsányi summarize the complexity of this situation:

The narrative of Central and Eastern Europe being bought up by China fits neatly with historical images of an Eastern Europe dominated by communism. In reality, it is the rest of Europe that is economically far more dependent on China and that has, too, put national interests above the common EU position. Pointing fingers at Central and Eastern Europe with the easily refutable claims of economic dependency distracts from the fact that Western European dealings with China have impacted the ability of the EU to form a common position. The intra-EU blame game can poison the relations among various European regions and is certainly not helping the EU in reaching unity on the issue.

(Hutt & Turcsányi, May 27, 2020)

Looking at China's share in exports (see Table 6.3), excessive dependence on China can be easily refuted in the case of the Visegrád countries. All three old EU member states have significantly higher China ratios in their exports, whereas Germany's dependence on China is outstanding. As far as Chinese investment is concerned in terms of GDP, Germany, Italy, and France show higher Chinese investment ratios than Poland and the Czech Republic (see Table 6.4). It must be added that Hungary's performance is above average in this aspect like that of the EU non-member states in the Balkans. But we can argue that Central European countries did not attract an excessive amount of Chinese investment which would undermine the unity of the EU in any way.

Table 6.3 China's share in the respective country's export (2018; percentage)

Germany	7.07
France	4.33
Italy	2.82
Slovakia	1.72
Hungary	1.91
Czech Republic	1.28
Poland	0.96

Source: Own compilation based on World Bank WITS database.

Table 6.4 Chinese FDI as percentage of GDP, ranking based on the relative size of Chinese FDI to GDP

	Chinese FDI stock between 2005 and 2020 (billion $)	GDP (billion $, 2019)	Chinese FDI as of GDP (%)
Hungary	5.88	163	3.60
Italy	26.75	2,004	1.33
Germany	47.58	3,861	1.23
France	28.48	2,716	1.05
Czech Republic	0.96	251	0.38
Poland	2.28	596	0.38

Source: Own calculation based on World Bank data and AEI's dataset "The China Global Investment Tracker." The dataset was updated in early 2020.

6.7 Challenges

In our assessment, China is currently not even close to challenging American leadership in technological development now, but is rapidly catching up in every segment; and there are two areas where China is on the verge of overtaking the United States: 5G and AI. In these two segments, the United States is doing its best to slow China's advance. There are also areas where China has a strategic advantage over the United States, such as the penetration of smartphone solutions, financial services on smartphones, and so on, in daily life and the acceptance of these solutions by the population.

The United States' containment strategy has proven successful, as most countries of the EU have refrained from cooperating with China, especially in building their 5G networks. At the same time, we have been able to show that the EU needs a different approach to China's technological progress than the one the United States has adopted for the following reasons:

- the EU does not claim global leadership, and it does not compete with China for geopolitical supremacy;
- the EU is less homogeneous than the United States in terms of economic development goals; and
- the EU is made up of nation states that have different political and economic development goals based on their level of economic development and political culture and even history.

It is important to stress that European technological sovereignty can only be achieved if Europeans rely on European technologies and companies in core areas. In other words, distancing ourselves from foreign technologies also means getting rid of American influence in this area. We have also come to the conclusion that there are no competitive European firms on a large scale to serve as European champions.

This problem can be traced back to the lack of a successful industrial policy at the EU level. From this point on, countries are on their own, or worse, they try to launch their industrial policies at the Single Market where typical business development tools are either banned or restricted. In our estimation, this situation will not change in the near future, keeping the EU, its countries, and companies in a technologically dependent situation. Countries like Hungary and Poland have realized that diversification is the key; other countries accept geopolitical arguments to reject Chinese technology and cooperate with China in this area.

In our opinion, there are several steps to dealing with this situation or at least easing the tensions:

1   There must be a wider scope for an interventionist policy approach at both the EU and country level to facilitate funds for technological development.

2   The available common funds for innovation, R&D need to be further increased at EU level.
3   Solutions must be found that take into account the different economic development needs of countries. This will avoid European countries being played off against each other (by other major powers).
4   The interventionist policy approach does not preclude more competition in the market. But if the aim is to create competition in the market, this must exclude geopolitical considerations.
5   If the area was too sensitive, the geopolitical risk assessment must include all major players, not only China and Russia, but also the United States, Japan, Korea, and so on.

In this book, we could see that the outcome of the technological competition between the United States and China is in many ways crucial for the future development of the world economy and world politics. Not only is global leadership at stake in this competition, but we also need to redefine in this process what we mean by (digital) privacy: where the accepted boundaries between public and private spheres are in the 21st century. We also need to rethink what economic development strategies best fit this new domain, where technological sovereignty may be the key to protecting sovereignty in the first place, and how Europeans can maneuver within this environment, as they seem to be at a disadvantage in this competition due to the institutional peculiarities of the EU. The European aspect of this competition can help us understand that European integration is not a goal *per se*, but a tool to achieve the goal of a more secure and prosperous future.

Notes

1 Ken Moak explains the key elements of the development strategy this way:

> Key to the 'internal circulation' was innovative manufacturing and increased private consumption. The former was to be realized through a massive spending of 1.4 trillion U.S. dollars over the next five years on innovation to climb the manufacturing value chain and become self-sufficient in advanced semiconductors and other technologies. The latter was to be attained by boosting household income through urbanization, turning migrant workers into city dwellers and enlarging the around 500 million middle-class population.
>
> (Moak, September 14, 2020)

2 The production of semiconductors is a case in which we can easily show why more self-sufficiency is required. In 2016, 16 percent of semiconductors were produced in China while the rest was mainly imported from the US (Kelly, September 3, 2020).
3 AEI demographics expert Nicholas Eberstadt draws our attention to the folly of underestimating the power of demography:

> Demographics may not be destiny, but for students of geopolitics, they come close. Although conventional measures of economic and military power often receive more attention, few factors influence the long-term competition

between great powers as much as changes in the size, capabilities, and characteristics of national populations.

(Eberstadt, 2019)

4 Between 0.9 and 1.3 percent of all inward FDI by the ultimate investing country.
5 The €4.9 billion in 2018, which was 1.3 percent of all inward FDI using the ultimate investing country approach.
6 The €1.6 billion in 2018, which was 6 percent of all outward FDI by the ultimate investing country.
7 According to the Hungarian Ministry of Foreign Affairs and Trade, about 15,000 people are employed in Chinese companies in Hungary and about $ 5 billion has been invested by Chinese companies in the last ten years.

References

Aron, L. (2019, April 4). Are Russia and China Really Forming an Alliance? The Evidence Is Less Than Impressive. *Foreign Affairs*, Retrieved from: https://www.foreignaffairs.com/articles/china/2019-04-04/are-russia-and-china-really-forming-alliance

Blanchette, J. & Polk, A. (2020, August 24). Dual Circulation and China's New Hedged Integration Strategy. *Center for Strategic and International Studies (CSIS)*, Retrieved from: https://www.csis.org/analysis/dual-circulation-and-chinas-new-hedged-integration-strategy

Cheng, E. (2020, November 4). China's Xi Says Country Will Speed Up Trade Talks with EU, Japan and South Korea. *CNBC*, Retrieved from: https://www.cnbc.com/2020/11/04/chinas-xi-says-country-will-speed-up-trade-talks-with-eu-japan-rok.html

Eberstadt, N. (2019, July/August). With Great Demographics Comes Great Power. Why Population Will Drive Geopolitics. *Foreign Affairs*, Retrieved from: https://www.foreignaffairs.com/articles/world/2019-06-11/great-demographics-comes-great-power

Guppy, D. & Sicong, X. (2020, October 25). Guide to China's Dual Circulation Economy. *CGTN*, Retrieved from: https://news.cgtn.com/news/2020-10-25/Guide-to-China-s-dual-circulation-economy-US8jtau4h2/index.html

Hutt, D. & Turcsányi, Q. R. (2020, May 27). No, China Has Not Bought Central and Eastern Europe. *Foreign Policy*, Retrieved from: https://foreignpolicy.com/2020/05/27/china-has-not-bought-central-eastern-europe/

Kelly, É. (2020, September 3). Decoding Europe's New Fascination with 'Tech Sovereignty'. *Science Business*, Retrieved from: https://sciencebusiness.net/news/decoding-europes-new-fascination-tech-sovereignty

Kratz, A., Huotari, M., Hanemann, T. & Arcesat, R. (2020, April 08). Chinese FDI in Europe: 2019 Update. *Rhodium Group (RHG) and MERICS*, Retrieved from: https://mimderics.org/en/report/chinese-fdi-europe-2019-update

Liechtentstein, M. (2020, November 18). China Prepares for a Storm. *Geopolitical Intelligence Services*, Retrieved from: https://www.gisreportsonline.com/china-prepares-for-a-storm,defense,3365.html

Mead, W. R. (2013, December 2). The End of History Ends. *The American Interest*, Retrieved from: https://www.the-american-interest.com/2013/12/02/2013-the-end-of-history-ends-2/

Moak, K. (2020, September 14). Why 'Dual Circulation' Policy Makes Sense. *CGTN*, Opinion, Retrieved from: https://news.cgtn.com/news/2020-09-14/Why-dual-circulation-policy-makes-sense-TLSR4osXfi/index.html

Pieke, F. N. (2019, August 27). Palpable Misconceptions and Possible Over-Reactions. *MERICS*, Retrieved from: https://merics.org/en/analysis/palpable-misconceptions-and-possible-over-reactions

Stern, J. (2020, October 13). Merkel's China Reset Is Mostly Likely Hollow. *Foreign Policy*, Retrieved from: https://foreignpolicy.com/2020/10/13/china-germany-united-states-reset-huawei/

US State of Department (2020, October 17). The Transatlantic Alliance Goes Clean. Fact Sheet. *Office of the Spokesperson*, Retrieved from: https://gr.usembassy.gov/the-transatlantic-alliance-goes-clean/

Wong, R. (2017). China's Rise: Making Sense of EU Responses. *The Journal of Contemporary China Studies, 2*(2), pp. 111–128.

Yao, K. (2020, September 15). What We Know about China's 'Dual Circulation' Economic Strategy. *Reuters*, Retrieved from: https://www.reuters.com/article/china-economy-transformation-explainer-idUSKBN2600B5

Yuzhu, W. (2017, September). Economic Rebalancing Should Be the Core of China's Supply-Side Reform. Part of Chapter 2. Macroeconomic Policy Coordination. In Parallel Perspectives on the Global Economic Order. *Center for Strategic and International Studies*, A U.S.-China Essay Collection, pp. 18–20.

Index

9780367652548